AIDS
and
Substance Abuse

ABOUT THE EDITOR

Larry Siegel, MD, is in the practice of Internal Medicine and Addictionology in Key West, Florida. His primary professional activities at present include care of people with AIDS and outpatient treatment of substance use disorders. Dr. Siegel co-founded the Delphos Alcohol and Drug Treatment Center in Key West and is currently chairperson, AIDS and Chemical Dependency Committee, American Medical Society on Alcohol and Other Drug Dependencies Inc., Chairperson, AIDS Task Force, National Association of Lesbian and Gay Alcoholism Professionals and Member, AIDS Task Force, American Association of Physicians for Human Rights. Dr. Siegel has given many workshops and lectures on the subject of AIDS and Chemical Dependency, and he organized the first and second Annual Forums on AIDS and Chemical Dependency in Ft. Lauderdale, Florida in February of 1987 and Phoenix, Arizona in February 1988.

AIDS
and
Substance Abuse

Larry Siegel
Editor

AIDS and Substance Abuse, edited by Larry Siegel, was simultaneously issued by The Haworth Press, Inc., under the title *Acquired Immunodeficiency Syndrome (AIDS) and Substance Abuse*, a special issue of *Advances in Alcohol & Substance Abuse*, Volume 7, Number 2 1987, Barry Stimmel, journal editor.

Harrington Park Press
New York • London

ISBN 0-918393-59-0

Published by

Harrington Park Press, Inc., 12 West 32 Street, New York, NY 10001
EUROSPAN/Harrington, 3 Henrietta Street, London WC2E 8LU England

Harrington Park Press, Inc., is a subsidiary of The Haworth Press, Inc., 12 West 32 Street, New York, New York 10001.

AIDS and Substance Abuse was originally published as *Advances in Alcohol & Substance Abuse*, Volume 7, Number 2 1987.

Cover design by Marshall Andrews.

Library of Congress Cataloging-in-Publication Data

Acquired immunodeficiency syndrome (AIDS) and substance abuse.
AIDS and substance abuse.

   "Simultaneously issued by The Haworth Press Inc., under the title: Acquired immunodeficiency syndrome (AIDS) and substance abuse a special issue of Advances in alcohol & substance abuse, volume 7, number 2, 1987."
   Include bibliographies.
   1. AIDS (Disease) 2. Substance abuse. I. Siegel, Larry, 1935-   . II. Title. [DNLM: 1. Acquired Immunodeficiency Syndrome. 2. Substance Abuse. WD 308 A18656]
RC607.A26A3415 1988   616.9'792   88-948
ISBN 0-918393-59-0

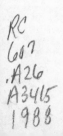

Contents

AIDS
and
Substance Abuse

EDITORIAL

The first national conference on Acquired Immunodeficiency Syndrome and chemical dependency was sponsored by AM-SAODD and held in Ft. Lauderdale, Florida in February 1987. A number of the country's experts gathered there to explore some of the relationships between the use of mood altering chemicals and the acquisition or progression of AIDS. While most experts are of the opinion that the human immunodeficiency virus (HIV) (formerly HTLVIII, LAV) is the cause of AIDS, some major questions regarding etiology remain unanswered. Among these questions are: Why is it that so few of the T cells that are destroyed in AIDS are infected by the virus? Why is the virus not recoverable from some patients who are infected or have AIDS? Why is the virus cytoxic in vitro in just a short time, but is latent and noncytoxic in vivo for years? And why is it that large percentages of certain populations such as groups of somewhat cloistered Venezuelan Indians show signs of infection by this virus, but do not develop AIDS? Clearly, many factors may play a role in AIDS. The goal at this conference was to explore the role of alcohol and other drugs in this epidemic, and to begin to look at what chemical dependency professionals can and should do.

The relationship between HIV and chemical dependency remains to be more clearly defined. Issues of concern include: (1) Can and do some drugs and alcohol alter the immune system? Specifically, what is their effect on T cell numbers and function? (2) Do drugs and alcohol play a role in predisposing to AIDS? (3) Why do some who are exposed to this virus get infected, and more critically, what prevents others, who are exposed, from being infected? (4) Why do so many who are infected with HIV *not* get AIDS? (5) What is there that is different about those who *do* get it, and those that don't, neurologically, psychiatrically, medically? (6) Do drugs and alcohol impact on behavior in such a way as to increase risk of primary infection with HIV or progression of disease once infection has occurred? (7) Are primary prevention of drug abuse or intervention in users useful strategies in containing this epidemic: If so, what do drug treatment professionals need to know to accomplish this? (8) What are the legal, moral and ethical issues surrounding AIDs for chemical dependency programs?

Many of these issues are explored in this issue of *Advances*. These papers however still leave more questions unanswered. Directions for future research are a little clearer. That alcohol and other drugs affect the immune system and our behavior is known. That we as chemical dependency professionals need to stay exquisitely "tuned in" seems evident. "More will be revealed." It is anticipated that this collection of papers will be but the first step in specifically addressing HIV infection and chemical dependency for those in the field.

Larry Siegel, MD
Chairman, AIDS and Chemical
Dependency Committee, AMSAODD
American Medical Society on Alcoholism
and Other Drug Dependencies

The Impact of AIDS on the Chemical Dependency Field

Robert G. Niven

This paper provides an overview of the effects the AIDS epidemic might have on the provision of services to alcohol and other drug dependent or abusing persons.

This review attempts to be comprehensive in scope and is primarily intended to raise issues for identification and further consideration: as such, it is necessarily limited in depth of consideration of individual issues.

An attempt is made to foresee some outcomes and since several of these would be adverse for the chemical dependency treatment field, they will be controversial, adding to the inherent controversy existing in our society as related to identification, treatment and prevention of AIDS.

THE GENERAL IMPACT OF AIDS ON SOCIETY

The spread of the organism responsible for the disease, Acquired Immune Deficiency Syndrome (AIDS) threatens to result in a worldwide epidemic of proportions witnessed by few (if any) living persons. Though the virus causing AIDS, the Human Immunodeficiency Virus or HIV (previously called LAV/HTLV-III) has apparently been in this country since 1978,[1] it has only been since 1981 that the morbidity associated with it has been recognized.[2,3] The virus itself was identified independently by research teams in France and in the United States, a discovery associated with one of

Robert G. Niven is Associate Professor of Psychiatry, Wayne State University School of Medicine.

3

the first controversies concerning the disease. The intense focus of both the health professions and the lay public has undoubtedly been secondary to the early recognition of the severe morbidity and high mortality of the disease associated with the lack of any substantive therapies. Initially the disease was thought to occur only in those in the so-called high risk groups (male homosexuals, intravenous drug abusers, transfusion recipients, Haitians, and infants of mothers infected during pregnancy) but increasingly, it is recognized to be spreading in the heterosexual population. This recognition, coupled with estimates of prevalence, are alarming to all who examine the data.[4,5]

The World Health Organization reports cases of AIDS in more than 100 countries and estimates that between 5 and 10 million persons worldwide are carriers of the virus and that up to 100 million may become infected in the next ten years. In the United States, over 50,000 cases have been reported by early 1988 and the Centers for Disease Control estimates 270,000 cases by the year 1991 with 179,000 deaths.[6] Since the disease frequently affects those in the prime of life and since it is expensive to treat, there are enormous financial implications for even the wealthiest of countries and the decimation of some countries is possible.[7]

In summary, presuming no cure is found in the near future, AIDS will become a disease of devastating proportions with adverse effects not only on those affected or at risk but, indirectly on all of society. It clearly will be the health care issue for the rest of this century. It will generate even more public concern than it already has and will force an examination or re-examination of a number of issues important to the alcohol and drug abuse field and indeed to all of medicine.

GENERAL IMPACT ON THE
CHEMICAL DEPENDENCY FIELD

There are several major ways in which AIDS related issues will affect the chemical dependency field and for the most part, such effects have much more potential to be adverse than beneficial.

First, our field functions as part of the overall health care delivery system and as such will be affected by changes that affect the entire system. AIDS, more than any other disease in modern history, will involve large segments of the public in making decisions concerning its prevention. Since many of these decisions involve potential conflict between the uninfected and the infected or between people in differing risk categories, AIDS might be considered a public policy disease with the potential for enormous conflict amongst large segments of society and these would impact the alcohol and drug treatment system as much as any other, if not more.

Since the spread of the virus (for the vast majority of those infected) involves some element of behavior over which they have at least some control and choice, AIDS will involve a greater societal and health focus on disorders involving risk-taking, impulse control, compulsivity and hedonism; this will include alcohol and drug taking behaviors. Such a focus will be stimulated by many forces. Prevention efforts currently and in the near future will rely heavily on strategies designed to reduce behaviors placing people at risk of contracting the virus.

Since intoxication with any drug, including alcohol, is associated with impaired judgement and at least in some cases, with increased risk-taking, clinical, research, prevention and public policy efforts aimed at better understanding and reducing these behaviors will increase. Intense public debate over the effect of traditional public health approaches to containment of contagious disease, which involve case identification and contact tracing, will challenge the principle of confidentiality which has been very important in the development of drug treatment programs.

Since intravenous drug abusers (IVDAs) make up one of the high risk groups for exposure to the HIV virus, considerable attention will be given this group. There will undoubtedly be renewed interest in Methadone Maintenance programs as a potential technique for minimizing exposure to the virus. There is already interest in providing sterile needles to IVDAs for similar reasons. Both of these developments will stimulate controversy in and outside of the drug abuse field.

The presence in alcohol and drug treatment units for persons in all of the AIDS risk groups (IVDAs, male homosexuals, transfusion recipients and those with multiple heterosexual contacts) will force virtually every treatment program to consider policies to deal with diagnostic, treatment and prevention issues. Not doing so will tend to reinforce denial, minimize opportunities for prevention (primary and secondary), and likely will be incompatible with mainstream medicine.

From a research perspective, there is both opportunity and risk from the AIDS crisis. Although there is general knowledge that alcohol and drug abusing persons are at increased risk for a variety of infectious diseases, there is relatively little known about the specific effects of alcohol or other drugs on the immune system.[8] In addition to altering behavior in ways which might increase risk of exposure to the HIV virus, alcohol (and possibly other drugs) might interact with the immune system in a variety of ways with positive, negative or neutral outcomes in different populations. Thus, while there is a great need for increased research in this area, there is the risk that any increased funding might come via diversion of funds from other equally necessary alcohol, drug, or mental health funding. Since most funding for AIDS research is concentrated on basic biological questions and, since there remains in some quarters, the view that alcohol and drug research isn't solid, it will take a concerted effort by all interested parties to elicit increased funding for such research and to ensure that it doesn't happen at the expense of other important alcohol and drug research or treatment efforts.

From the medico-legal standpoint, there are few issues unique to the chemical dependency field but the field will be affected by virtually all medico-legal developments which promise to be many, varied and ongoing. AIDS has already had a major impact on public health law[9] and I believe this is minor compared to what may happen in the next few years. The list of potential lawsuits surrounding AIDS issues is almost limitless. All treatment programs will have to keep abreast of current and pending legislation, and civil and criminal AIDS related cases at both the federal[10] and state levels if they wish to remain in compliance with the relevant laws and minimize risk of being sued and of losing (there is no way to eliminate such risk). Since many of the legal issues will involve confidentiality and/or stigma, both extremely important to alcohol and drug patients and providers, it is essential that all providers become involved in professional societies formulating or seeking to influence AIDS related laws to minimize abuses of confidentiality statutes and to avoid stigmatizing those with or at risk for AIDS. For the individual practitioner, practicing state of the art medicine carefully individualized to meet the situation and needs of each patient with careful and complete documentation will provide the greatest combination of medical benefit at minimal legal risk.

SCREENING FOR HIV
IN TREATMENT FACILITIES

Screening for exposure to the HIV antibody is highly controversial at present, in society generally and in substance abuse programs specifically.[11,12,13] Positions range from proponents of universal mandatory screening to opponents of any screening.

Proponents of screening believe it will be helpful in preventing spread of the virus and by implication that specific knowledge of antibody status (in contrast to general knowledge of the issues) will be helpful either to the individual and/or to society in minimizing the number and/or impact of AIDS victims. Opponents of screening contest these beliefs and seem to fear screening will be used to stigmatize HIV positive individuals and/or those in known risk groups. There is incomplete scientific evidence concerning the impact of knowledge on changing behavior but I believe that most people accept the proposition that knowledge of antibody status is more helpful than harmful, which may be why most of the public supports screening.[14] Those who say a positive test is meaningless appear to ignore what a true positive (compared to a true negative) test means concerning risk of developing AIDS and to underestimate the potential benefit of a true negative test to those who are somewhat aware of their risk status and are seeking reassurance. Further, some proponents at both extremes may be using their positions to further aims other than dealing with AIDS (i.e., stigmatizing versus legitimizing homosexual behaviors). Proponents of universal and/or mandatory testing appear to overvalue potential benefits of screening and undervalue the problems associated with this approach while opponents of any screening may exaggerate the problems (primarily stigma and discrimination) sometimes associated with screening.

A complete analysis of all the issues surrounding screening is beyond the scope of this paper but the complexity and seriousness of this procedure at a programmatic and individual level demand that providers seeing persons at risk for exposure to HIV be up to date in their knowledge and considerate of potential positive and negative effects of either positive or negative tests (see Tables 1 and 2).

I believe widespread screening for the HIV antibody (or for the HIV itself if this becomes widely and inexpensively available) is inevitable for the following reasons. First, there are large numbers of people in all known risk groups who will want to have the test done, even when they understand the possible negative implications. Secondly, there are clinical situations in which testing will be

TABLE ONE

Potential Benefits of Screening

-Decreased anxiety of those in risk groups who are worried and who test (true) negative.

-Increased motivation of those who test (true) positive to engage in healthy behaviors including secondary and tertiary prevention for self and primary prevention re: others (i.e. safe sex).

-Increased time for those who test (true) positive to "get their affairs in order" and to respond both to decreasing health and new treatments which will emerge.

-Screening procedure may add to impact of general discussion with patients and enhance healthy behaviors.

-Screening is consistent with treatment of other health problems and may contribute to decreasing stigma associated with AIDS.

-Will enhance data base for research studies.

TABLE TWO

Potential Negative Effects of Screening

-Error in testing or interpretations by staff, patients
or others resulting in false diagnosis (either positive or
negative).

-Adverse response by patients or others (suicide,
aggressive and/or sexual acting out, discrimination), in
response to either positive or negative tests.

potentially helpful in differential diagnosis or in preventing transmission (i.e., blood transfusion or organ donation). Thirdly, there
are public health and medicolegal reasons for testing certain individuals in accord with prevailing statutes, for epidemiologic research, for determining legal questions of responsibility for transmitting the virus and ultimately, for contact tracing. Fourthly,
screening for employment in certain positions or institutions (already in place for the armed services) will increase and will likely
affect the medical profession (i.e., phlebotomists, surgeons, endo-
scoptists), with screening not necessarily being related to purely
scientific issues.[15] Lastly, screening for insurability for health and
life insurance will increase since risk assessment is a critical principle for insurers and ultimately, for those who pay the premiums.

In addition to the issues raised by screening itself, the confidentiality of the results raises additional problems, which may have
clinical, ethical and medico-legal aspects and might be expected
with greater frequency on treatment units than in general medical
units. Inappropriate fear in treatment unit personnel can lead to behaviors which threaten confidentiality. The duty to warn others of
dangers presented by a HIV positive patient will at times put staff in
direct conflict with confidentiality principles and laws.[16] The need
to provide third party payors with information to justify hospitalization may place reimbursement in conflict with confidentiality. Confidentiality never has been an absolute guarantee in medicine but
remains an important concept in facilitating treatment. It should be

maintained in dealing with AIDS issues whenever possible but the complexity and seriousness of the AIDS epidemic almost guarantees that this will remain one of the most troublesome areas for alcohol/drug treatment providers' and patients' years to come.

DRUG DEPENDENCE TREATMENT
FOR PERSONS WITH AIDS

Treating alcohol or other types of drug abuse or dependence in persons with AIDS (PWAs) or those at risk for it raises several problems which must be dealt with by all concerned with treatment. First, there is a need to increase our prevention efforts in treatment programs since we see large numbers of persons at risk for but who do not now have AIDS or the HIV.

Secondly, there is a need for increased treatment availability and efficacy for persons with alcohol/drug problems who are in high risk groups but do not have AIDS.

Thirdly, there is need for treatment for persons with chemical dependency who already have the virus and for those with the disease.

Few would argue with the first two which can be addressed in most treatment facilities, whether in- or out-patient, with relatively little in the way of program modification and with minimal increase in budget and staff. Several issues arise in treating AIDS patients in existing chemical dependence programs.

AIDS is a uniformly fatal disease with multiple physical and emotional complications. While some AIDS patients without current acute illness can be treated without any problem, others have medical (including psychiatric) needs which cannot be readily met in many treatment programs.[17] The presence of an AIDS patient on a treatment unit may evoke a variety of reactions from other patients and from staff requiring much skill and time to deal with in a manner which facilitates the treatment of all. Having AIDS, like having many other fatal diseases, raises treatment issues which are quite different from those in most chemical dependent patients, for whom death is much more remote (the vast majority in most units). One such issue involves the extent to which abstinence from psychoactive drugs is possible or practical. Another is the extent to which issues of dying are dealt with in the program. Arranging appropriate after-care, a difficult enough task with many patients becomes even more complicated with PWAs, although this problem should diminish as support groups, facilities and programs become more available and experienced. Confidentiality and anonymity are other obvious issues faced by the patient and program. Fiscal and ethical considerations will undoubtedly also be raised regarding chemical dependency treatment as they will with other types of treatments for AIDS, the essential question being, "To what extent should resources be spent on people with a fatal disease, particularly if they 'brought it on themselves'?"

I conclude that PWAs and chemical dependency should indeed be treated whenever it is possible to do so. Many such persons, particularly those in the earlier stages of disease may benefit from recovery programs and I believe the self-help programs utilizing the principles of Alcoholics Anonymous will prove beneficial in helping such persons deal with their AIDS as well as with their chemical dependency. Some patients however, mostly those with current complications and/or very advanced disease (particularly those with organic brain syndromes) will require modification of standard approaches and some will not benefit from participation in standard programs (and some could conceivably even be harmed). Even those who can and do benefit likely will need more help then they can get in most programs and the emergence of special programs,[18] including those exclusively for PWAs will likely occur in larger urban centers. The development of alcohol/drug programming in both in- and out-patient AIDS clinics is desirable from a prevention and treatment standpoint.

The decision to admit PWAs to a treatment program should not be made without regard to the above issues, and should be individualized as with any other patient. Ideally, the patient should be admitted only if treatment can be provided competently and professionally and with the expectation of a positive outcome equal to that in other readily available programs in which the PWA might participate. The HIV antibody screening test should not be used to screen out positive testers from treatment programs. Treatment staff also have an obligation to make effective referrals for any patient who is determined to be inappropriate for treatment in their facility.

CONCLUSIONS AND RECOMMENDATIONS

1. There will be few if any programs that will not be affected in some way by the AIDS epidemic, and all alcohol/drug programs need to maintain an up-to-date awareness of issues which may affect their patients or program.

2. All patients entering programs should be screened, by sensitive thorough history taking, for risk of exposure to the HIV virus, keeping in mind that heterosexual spread of the virus is increasing. I believe that all persons at risk of exposure whose antibody status is not known, should be screened with the HIV antibody test, on a voluntary basis, unless doing so poses some direct risk which outweighs the value of the test. At the very least all programs need to have a policy concerning screening. All patients, whether they accept or decline the test should have their risk status explained to them, carefully and repetitively (in order to overcome defenses).

3. All programs should have educational components designed to help their patients and significant others minimize risk of future exposure to the HIV virus. Alcohol/drug programs in this country have considerable expertise in dealing with such difficult issues and in understanding the defenses people have which enable continuation of risk-taking behaviors. Since we see many people at risk for contacting the HIV virus, and therefore have an opportunity to make a significant contribution to slowing the spread of this disease, our field cannot ignore such prevention opportunities.

4. The chemical dependency field, in particular the major organizations in it, need to develop committees to deal with fiscal and public policy issues relevant to AIDS. In particular we need to be active in supporting research on AIDS making sure that this does not happen at the expense of other important research efforts. Additional issues which will require ongoing attention pertain to confidentiality and to funding of adequate treatment programs for PWA's. Failure to participate in the political activities will likely result in fiscal or policy decisions which negatively impact on some

of the patients we treat or some of the gains we have made in the past decade.

5. The AIDS problem has the potential to resurrect some of the old conflicts in the chemical dependency field (i.e., alcohol-vs-drugs) and to pit certain groups against others (i.e., homosexual-vs-straight). While there is much legitimate controversy in our field we should all strive to maintain a focus on solving the problems presented by the HIV virus, while maintaining and improving treatment for all alcohol and drug dependent persons. It is imperative we not let this problem fragment our field further than it already is. Achieving this goal will require the active effort of most of the organizations and their members engaging in long term constructive programs to resolve the myriad of issues raised by the AIDS epidemic.

NOTES

1. Selwyn, PA. AIDS: What is Now Known. HP Publishing Co., 10 Aster Place, New York, NY 10003.

2. CDC: Pneumocystis pneumonia. Los Angeles: MMWR, 1981; 30:250-252.

3. CDC: Kaposi's Sarcoma and Pneumocystis Pneumonia Among Homosexual Men. New York City & California: MMWR, 1981; 30:305-308.

4. Heterosexuals and AIDS. The Atlantic Monthly, Feb. 1981:39-58.

5. The Big Chill: Fear of AIDS. Time, Feb. 16, 1987:50-53.

6. CDC: Update: Acquired Immunodeficiency Syndrome — United States. MMWR, 1986; Vol. 35:49-50.

7. Africa in The Plague Years. Newsweek, Nov. 24, 1986:44-47.

8. MacGregor RR. Alcohol and Immune Defense. JAMA, 1986; 256:1474-79.

9. Matthews GW, Neslund US. The Initial Impact of AIDS on Public Health Law in The United States. JAMA, Jan. 16, 1986; Vol 257:#3:344-352.

10. Pascal CB. Legal Issues for AIDS and Drug Abuse Treatment Programs. Office of General Counsel, Public Health Service, Dept. of Health & Human Services, June 1986.

11. Levine C, Bermel J. AIDS: Public Health and Civil Liberties. Hastings Center Report, Dec. 1986.

12. Alvig OV. Mandatory Screening for HIV Antibody Tester to Editor. JAMA, Feb. 6, 1987; Vol. 257:#5:625.

13. Proceedings of Symposium on AIDS and Chemical Dependency. NIAAA, April 1986:21-29.

14. Gallup Poll. Newsweek, Nov. 24, 1986:35.

15. AIDS Job Bias Growing Fast in Health Industry. American Medical News, Feb. 27, 1987:34.

16. Binder RL. AIDS Antibody Tests on Inpatient Psychiatric Units. Am J Psychiatry, Feb. 1987:144:2.

17. Dilley JW et al. Findings in Psychiatric Consultations with Patients with AIDS.

18. Cohen MA, Weisman HW. A Biopsychosocial approach to AIDS. Psychosomatics, April 1986; Vol. 27:#4.

Should There Be HIV Testing in Chemical Dependency Treatment Programs?

William Hawthorne, MD
Larry Siegel, MD

The AIDS epidemic will have a major impact on alcohol and substance abuse treatment. As the epidemic spreads more people will be affected. An increasing number of people infected with the virus will seek chemical dependency treatment. The association between the diseases of chemical dependency and AIDS is intense. Aside from the obvious association between IV drug use and AIDS, the fast track lifestyle is often shared between people at risk for AIDS and chemical dependency. Guidelines for caring for HIV positive patients in substance abuse treatment programs are now available.[1] Use of mood altering chemicals can predispose patients to risky behavior which may lead to exposure to the virus. A not so well known association is that many mood altering chemicals are speculated to be cofactors, increasing susceptibility to the causative agent (widely believed to be the Human Immunodeficiency Virus, HIV) and aiding in the progression of the disease.[2] This may be because many mood altering chemicals including alcohol are, like AIDS, immunosuppressive.

Chemical dependency treatment providers will be increasingly affected by the AIDS epidemic. Policies and procedures are required to assist clinicians to respond to the special needs of these patients. Educational programs for staff, patients and families can provide accurate information to deal with real and imagined anxieties. A frequent and controversial concern is the issue of testing patients for the presence of the HIV antibody.[3,4,5,6] The HIV antibody test was developed originally to protect the blood supply and make transfusions safe. It has subsequently been widely utilized as a diagnostic tool for AIDS.

While AIDS, thus far, is most often a fatal disease, some patients have survived for many years. We are not in a position to predict what percentage of patients infected with HIV go on to develop full blown AIDS. We know it is a significant percentage. A patient struggling with early recovery who finds out that he is HIV positive may "give up" on life. A patient who is HIV negative may be falsely reassured and have his sense of invulnerability reinforced. We now know that there are both false positives and false negatives. A person may be negative because they have been infected but have not yet had time to develop antibodies or may be unable to do so.

A number of experts and the Centers for Disease Control have suggested that all patients admitted to chemical dependency facilities should be tested for the presence of the HIV antibody. It is reasoned that a person should know if they have been infected by the virus and have some risk of developing AIDS.[3,7] It is reasoned that since most, if not all of these HIV positive people are infectious, a positive test would encourage safe sex practices and discourage needle sharing. While there is good data suggesting that education is associated with altered sexual practice there is little data to support the notion that knowledge of HIV antibody status does (see below).

The model for this type of preventive effort comes from another sexually transmitted disease (STD), syphilis. With syphilis, routine mandatory testing is required in some states in a variety of circumstances including applying for a marriage license and entering a hospital. If a person has a positive test for syphilis, that person and all sexual contacts are tested and treated. This "contact tracing" is dependent on information supplied by the propositus. This model has been effective in controlling the spread of syphilis. However, in the instance of HIV positivity occurring mainly in homosexual men, IV drug addicts and prostitutes, the ability to do contact tracing is very limited. The opportunity for intervention with treatment does not exist, and the capacity to further limit HIV spread is already markedly limited. The model therefore breaks down when applied to AIDS and raises questions about whether the HIV test should be routinely offered in a chemical dependency setting. Since there is currently no treatment available for antibody positive individuals and because the psychological and social costs are high, it is at least questionable whether routine testing is in the best interests of our patients.[5,9] Confidentiality is not always protected in many jurisdictions.

There are clear social costs to pay for being HIV positive.[8,10] Insurance companies are denying health and life insurance to people who are HIV positive. The Justice Department has suggested that federal protection against job discrimination may not extend to people who are HIV positive. A number of federal agencies including the armed forces and State Department have announced their intention to discriminate in hiring people who are HIV positive. People have lost their jobs when it became known that they were HIV positive. Some have expressed concern that if coercive testing is instituted and the sanctions become severe enough, potentially positive persons might go underground and avoid testing altogether.[2]

There is an axiom in medicine, "Doctor, do no harm." With so little to gain and so much to lose by the HIV test, one must be cautious in recommending to a patient to take the test. We cannot become servants to a broader public health objective which may result in more harm to our patients. The public health objective is to prevent the spread of the disease. It is not yet to benefit the individual patient. Yet this public health objective may not be successful because there is, as yet, little evidence to support the contention that knowledge of HIV status impacts on sexual behavior. On the contrary, Ostrow[11] has reported that information on HIV status does not further alter sexual behavior in a cohort of homosexual and bisexual men. Further, increasing reports of attempted and actual suicide by individuals after being informed of HIV status are deeply troubling.[11,12] When the test is done, very careful pre and post test counseling must be included as part of the test procedure.[13]

More importantly, safer sex practice and no needle sharing should be practiced by everyone to prevent a whole variety of diseases, of which AIDS is the most serious. Such practices should not be confined to just high risk groups or individuals who are HIV positive. The disease has spread beyond risk groups and it may be too late once an individual is HIV positive. It is just as easy to argue that a chemically dependent person who learns he is HIV positive will "give up on life" and continue to drink, drug and spread AIDS, as it is to argue that he will recover and modify his behavior to prevent the spread of AIDS.

There are many ethical and legal issues that remain unanswered.[4] What if a bisexual man finds out he is HIV positive but refuses to tell his wife who does not know he has sex with men? What is the "duty to warn" the wife? There are no easy answers to this type of question. Expected pregnancy is probably one area where the need for HIV testing is becoming clear. Patients in high risk groups who are anticipating having children should be encouraged to receive the test since there is a high incidence of AIDS among their children.

What if we learn that there are specific things that people who are HIV positive can do for themselves to prevent progression? Perhaps there will be an intervention discovered that will prevent the development of AIDS among people who are HIV positive. Perhaps it will be proven that significant behavioral changes will alter the progression of the disease. If this happens there will then be help that can be offered to the HIV positive patient and rationale for more strongly suggesting the test. But until that time, routinely ordering the test does not seem to be beneficial.

What kind of an HIV testing policy should a program have? (1) Do not require routine testing for all patients. (2) Provide for voluntary testing at alternative test sites which will protect patient confidentiality. (3) Leave it up to the patient whether to tell staff or fellow patients. (4) Provide for pre and post test counseling which will address the fears and expectations surrounding the test. Most patients do not just want the test: They want a negative result! (5) Provide education to staff, patients, and families about the disease, with up-to-date knowledge in this rapidly changing field. The patient should be instructed about safer sex practices and the dangers of needle sharing. Confidentiality, while always important, is especially important with AIDS.

NOTES

1. Guidelines for facilities treating chemically dependent patients at risk for AIDS or infected by HIV virus. American Medical Society on Alcoholism and Other Drug Dependencies, Inc. 6525 W. North Avenue Suite 204, Oak Park, Illinois 60302.

2. Siegel L. AIDS: Relationship to alcohol and other drugs. Journal of Substance Abuse Treatment 1986 3:271-274.

3. Bayer R, Levine M, Wolfe S. HIV antibody screening. Journal of the American Medical Association 1986 256:13;1768-1774.

4. Ginzburg M, Gostin L. Legal and ethical issues associated with HTLV-III diseases. Psychiatric Annals. March 1987 16:3;180-185.

5. Siegel L. Blanket AIDS testing counterproductive. U.S. Journal September 1986.

6. The pros and cons of the HTLV-III antibody test. Position Paper by Bay Area Physicians for Human Rights. May 1986.

7. Johnson E, Perez G, Slim J. AIDS tests in office practice. Medical Aspects of Human Sexuality. June 1986 18, 24-25, 29, 32.

8. Leishman K. Heterosexuals and AIDS. The Atlantic February 1987 39-58.

9. Chase M. Asking AIDS victims to name past partners stirs debate on privacy. The Wall Street Journal 29 January, 1987 1, 16.

10. Binder R. AIDS antibody tests on inpatient psychiatric units. American Journal of Psychiatry February 1987 144:2;176-181.

11. Ostrow D. Antibody testing won't cut risky behaviour. AMA News 5 June 1987.

12. Seckinger D. Abbott National Teleconference on HIV Testing. 18 June 1987.

13. Smith T. Counseling gay men about substance abuse and AIDS. National Institute on Alcohol Abuse and Alcoholism Pamphlet Acquired Immune Deficiency Syndrome and Chemical Dependency, DHHS Publication No. (ADM) 87-1513 1987;53-59.

To Test or Not to Test: The Value of Routine Testing for Antibodies to the Human Immunodeficiency Virus (HIV)

Barry Stimmel, MD

Of all the controversies in medicine, and there are indeed many, none has aroused more emotion, often accompanied by more heat than light, than the subject of routine testing for infection with the Human Immunodeficiency Virus (HIV). Among professional organizations dedicated to studying this issue, consensus has not been obtained, even with respect to routine testing within groups whose activities put them at a marked increased risk for developing HIV infections, such as intravenous drug users.

The Public Health Service has clearly recommended that intravenous drug abusers be routinely counseled and tested for HIV antibody.[1] Other equally prestigious organizations, such as the American College of Physicians, presumably reviewing the same data base, concluded that testing in and of itself does not provide sound prognostic information for asymptomatic persons and that "screening of individuals at high risk [should] be conducted, if at all, on a case-by-case basis."[2] The complexity of this issue can perhaps be best demonstrated by the recommendations in a report of the Institute of Medicine-National Academy of Sciences. These organizations find mandatory screening of at-risk individuals to be not ethically acceptable, yet recommend compulsory measures "with full due process protection" for "a recalcitrant individual who refuses repeatedly to desist from dangerous conduct in the spread of infection."[3] How such an individual is to be identified in the absence of mandatory testing or compulsory reporting remains to be defined.

Contributors to this issue of *Advances* are no different with respect to their participation in this controversy; routine testing has both its advocates and its opponents whose views appear consistently throughout the issue. In an attempt to better delineate the issues at hand, the *problems* associated with testing, which have resulted in an inability to achieve consensus, will be defined (Table 1).

ABILITY OF TESTING FOR HIV ANTIBODIES TO PROVIDE A SATISFACTORY LEVEL OF CONFIDENCE WITH RESPECT TO HIV INFECTION

Of the three techniques available to document the presence of HIV infection (Table 1), the only method relevant to routine screening is detection of HIV-specific antibodies produced by an individual's immune system in response to the presence of the virus. The test most frequently used is the ELISA, the enzyme-linked immunoabsorbent assay, with subsequent confirmation by the Western blot assay method. Although the accuracy of the ELISA to determine HIV infection has been listed as being as high as 99.6% sensitivity and 99.2% specificity, in fact when one reviews data based on individuals at high, medium, or low risk, somewhat less compelling results are obtained. The sensitivity of the ELISA is directly dependent on the cutoff value set by the manufacturers of the test kits. The ELISA reacts to the presence of antibodies in the blood through development of a color reaction quantitated by a spectrophotometer. The higher the antibody level, the stronger the color. If the cutoff value is set at a low point, relatively faint specimens are recorded as positive, increasing the chance of detecting HIV with low levels of antibody, as well as the chances of a false positive reaction. In low-risk populations false positive rates have varied from 0 to as high as 50% with variations in rates noted even among different batches in a single manufacturer's kit.[3] False positive rates, however, can be significantly diminished to as low as 5 in

Table 1:

ROUTINE SCREENING FOR ANTIBODIES TO THE HIV

ADVOCATES	CRITICS
High degree of accuracy for HIV infectivity	Sensitivity and specificity varies with manufacturer and population studied
Can serve as reinforcer to modify high-risk behavior	High-risk behaviors should be eliminated regardless of reactivity
Can reassure repeatedly negative patients they do not harbor HIV	Those testing positive may become depressed and engage in self-destructive behavior
Can allow physician to make informed decision concerning treatment for changes in health status	Physician should assume all intravenous drug users to be potentially positive and act accordingly
Adequate measures for confidentiality can be established	Confidentiality can never be adequately assured
Allows for tracing of contacts	Ability to do contact tracing limited and fruitless as no prevention or cure for AIDS exists

100,000 if sequential tests are done by ELISA and then confirmed by Western Blot.[4]

In high-risk populations such as intravenous drug abusers, the sensitivity of the ELISA is quite good (99.5%). However, as many as 2.8% of those with a negative test may be infected, and as many as 7% of infected individuals may not be detected.[5] False negative tests are therefore a major concern in screening a high-risk population. In addition, depending on the kit used and the accuracy of the laboratory, a considerably smaller number of individuals with positive ELISA tests will be found to have positive Western Blot reactions.[6] The chances of a false positive ELISA are therefore higher in drug abusers than in low-risk populations.

The meaning of a positive and/or negative test is also not clearly agreed upon. Although it is felt that an individual may test "negative," yet carry the virus and be potentially infectious to others, it had been agreed that the time from exposure to HIV to production of measurable antibody was no longer than several months. A negative test, if repeated within a three- to six-month period, in the absence of further exposure to the HIV, was believed to provide a relative degree of assurance that indeed HIV infection had not occurred. Unfortunately, more recently acquired data suggest a much longer latency period precedes sero conversion. Although it has been questioned whether all individuals with antibody are actually infectious, and whether infectivity may be transient, it is generally agreed that those individuals with confirmed positive test results have the virus in their bloodstreams and are potentially infective.

THE NEED TO KNOW

One of the most frequently voiced arguments against routine testing for antibodies to HIV is that knowledge of antibody positivity is relatively useless, as nothing can be done to prevent the development of the disease and there is no known therapeutic cure. Other arguments are that, since individuals at high risk need to modify their behaviors, knowledge of a positive test is relatively unimportant and such knowledge will needlessly worry an individual and may even result in suicidal behavior. This logic is similar to that used by physicians several decades ago to avoid telling their patients that they had cancer.

Knowledge that an individual is negative may well serve as a motivating factor to encourage positive changes in behavior which places one or one's family at risk. Although anecdotal evidence suggests that knowledge of HIV status does not alter sexual activity, and in fact may result in an increase in self-destructive behaviors,[7] data also exist that the opposite can occur. Farthing et al. in a study of 324 gay men offered testing found 87% to agree, with only four individuals not wishing to know the results.[8] Although 65% had already modified their behavior, 93% thought they would be more likely to do so if shown to be positive, with 79% doing so if the test was negative. Three months after testing, half of those practicing safer sex felt that they were doing so as a result of the testing; half as a result mainly of counseling. The authors concluded that the overwhelming majority of men wanted to know their antibody status and that having the test did encourage safer sex practices.

Knowledge of HIV positivity may also be quite important with respect to an individual's personal health needs and care. For women, knowledge of antibody status may result in delaying pregnancy until more is known about the risk of transmission, as well as the need to avoid breastfeeding. Those individuals who are tuberculin reactive and HIV-antibody positive may need to have prophylactic chemotherapy to diminish the risk of reactivation of old tuberculin infections. Physicians who know a patient carries the HIV antibody will be more alert to the importance of: (1) evaluating nonspecific medical symptoms, (2) immediate treatment of existing infections, and (3) avoiding drugs which may cause immunosuppression. Although it is argued that physicians should be alert to these problems in any individual who is an intravenous drug abuser, at times the fact of intravenous drug abuse is kept from the physician.

The importance of a positive reactor's significant others knowing HIV status also cannot be underestimated. Evidence of a positive test may well serve to break through the denial experienced by many, to insure that appropriate precautions be taken with those with whom one is most intimate.

CONFIDENTIALITY

The issue of confidentiality is indeed the most compelling argument mustered against routine testing. There is no question that, regardless of the limited ability of the HIV to be transmitted to individuals other than through intimate contact or contamination with blood products, the widespread fear over developing AIDS from an individual who is antibody-positive cannot be underestimated.

A recent Gallup poll documented 52% of Americans favoring routine testing for all Americans with 90% favoring testing for immigrants, 88% inmates of Federal prisons, and 80% couples applying for marriage licences.[9] Of interest is the observation that education had a marked effect in determining the response. Only 39% of college graduates were in favor of all-inclusive testing as compared to 64% of respondents without high-school degrees. Testing of individuals with the potential of transmitting the virus was also markedly favored, with 80% wanting all medical personnel to be tested, 86% wanting the results made public and 57% feeling AIDS-infected workers should be prevented from treating patients.[10]

Insurance companies have already tried, on an individual basis, to prevent those individuals at risk for AIDS or known to be HIV-positive, from obtaining health insurance. Several states have introduced bills to mandate testing. In the Illinois Legislature, in 1987, at least 57 bills dealing with this issue have been introduced, and many have passed. These bills include requiring local health departments to report to school officials the identities of children affected with AIDS; making it a felony for an AIDS carrier to donate blood; and isolating AIDS virus carriers who continue to engage in acts that could transmit AIDS.[11] The armed forces has already instituted mandatory screening of all recruits with the discharge of those found to test positive. A number of other Federal agencies are also considering mandatory testing.

This fear is not confined to the United States, but indeed has "infected" other countries as well. The policy has probably been carried to its extreme in the West German state of Bavaria, which recently introduced regulations requiring compulsory screening of all those applying for public sector jobs and of non-Europeans seeking residency permits, and requiring forced isolation of those carrying antibodies to the AIDS virus who refuse to follow specified preventive measures.[12]

Unfortunately, even in the medical profession AIDS patients are not infrequently stigmatized. Residency programs which have a large proportion of AIDS patients among their inpatient service have been expressing concern over the effect that this has had over the recruitment of house staff. Kelly et al., in a study of 157 physicians asked to complete an attitudinal survey on the basis of two identical patient vignettes, with one patient identified as having AIDS and the other Leukemia, and sexual preference being either heterosexual or homosexual, found harsh attitudinal judgements associated with the AIDS portrayal.[13] This included much less willingness to enter into even a routine conversation when the patient's illness was identified as AIDS. If this disorder can elicit negativity and avoidance on the part of physicians—those most tuned to the presence of disease—then the fears concerning violation of confidentiality with routine testing certainly are based on reality.

THE ROLE OF COUNSELING

The one area in which there seems to be universal agreement is the need for an intensive, effective, and broad-based counseling program to educate persons concerning the nature of HIV testing as well as the significance of the results that are obtained. This counseling must be available to all intravenous drug users regardless of whether testing is requested. It should be comprehensive, addressing issues of modifying life behavior, symptoms associated with suppression of the body's immune response, and the dangers of continuing to engage in illicit drug abuse due to its effects on the immune system.

At present, consensus among physicians treating intravenous drug users in chemical dependency programs cannot be reached regarding the desirability of routine testing. This lack of consensus is understandable and not of extreme concern, providing that an individual physician's bias is not transmitted to the patient. The patient must be allowed the opportunity to become informed of the facts and to make a reasonable decision as to whether or not to be tested. It is unfortunate that in many programs the bias of the staff is reflected in the services available. This is regrettable since more than in other health settings there is frequent contact between health care worker and patient. Each contact provides an opportunity to educate which should not be lost.

The risk of intravenous drug abusers developing AIDS is great. Equally important is their potential to transmit the HIV to others through high-risk behavior. Anything that can be done to alleviate an individual's anxiety or to allow him/her to be better informed as to the nature of this disease and his or her risk must be accomplished. Not only do clients in drug programs need to be informed so as to decide whether the knowledge of their antibody status would be helpful or harmful but in addition outreach programs should be established to increase intravenous drug users' knowledge of prevention of HIV infection as well as encouraging them to enter treatment. For a physician to do less than provide this information, in a clear, open way, is to negate one's obligation to the profession and to society as a whole.

NOTES

1. Additional recommendations to reduce sexual and drug-abuse-related transmission of HTLV III/LAV. MMWR 1987 36:509-13.

2. Acquired immunodeficiency syndrome. Health and Public Policy Committee of the American College of Physicians and the Infectious Diseases Society of America. Ann. Int. Med. 1986 104:575-585.

3. Confronting AIDS: Directions for public health, health care and research. Committee on a National Strategy for AIDS, Institute of Medicine, National Academy of Sciences. National Academy Press, Washington, DC 1986 pp. 129, 130.

4. Meyer KB, Pauker SA. Screening for HIV: can we afford the false positive rate? New Eng. J. Med. 1987 317:238-241.

5. Barry MJ, Cleary PD, Fineberg HV. Screening for HIV infection: risks, benefits and the burden of proof. Law, Medicine and Health Care 1986 14:259-287.

6. D'Aguila R et al. Prevalence of HTLV-III infection among New Haven, Connecticut parenteral drug abusers in 1982-83. New Eng. J. Med. 1986 314:117-118.

7. Ostrow D. Antibody testing won't cut risky behaviour. AMA News 5 June 1987.

8. Farthing CF, Jesson W, Taylor HC, Lawrence AG, Gazzard BG. The HIV antibody test: influence in sexual behavior of homosexual men. National Institutes of Health AIDS Conference June 1987 p. 5 M 6.4 Bethesda, MD.

9. Most support widespread AIDS testing, Gallup reports. N.Y. Times Monday July 13, 1987.

10. Americans strongly favor AIDS testing for health workers. Health Care Competition Week. 13 July 1987.

11. Illinois Legislature sees some 57 bills on AIDS. American Medical News 5 June 1987.

12. Dickenson D. Bavaria requires AIDS testing. Science 29 May 1987 1057.

13. Kelly JA, St. Lawrence JS, Smith S, Hood HV, Cook DJ. Stigmatization of AIDS patients by physicians. Am. J. Pub. Health 1987 77:789-791.

AIDS Update – 1987

Donald I. Abrams, MD

SUMMARY. Human immunodeficiency virus (HIV) infection produces a spectrum of clinical syndromes, progressing in severity from asymptomatic infection through the life-threatening diseases of the acquired immunodeficiency syndrome (AIDS). Current knowledge about the epidemiology, virology, and clinical manifestations of HIV infection and AIDS are reviewed.

Over the past few years, a barrage of news coverage about the acquired immunodeficiency syndrome (AIDS) has given witness to the impact of this disease on our society and its devastating toll in young lives. In the United States, AIDS has become the eleventh most common cause of premature death. In San Francisco and New York City, AIDS now surpasses the cumulative impact of accidents, homicide, suicide, and cancer as the leading cause of years of potential life lost among single men aged 25 to 44 years. To respond appropriately and effectively to this epidemic, health professionals in many fields need to stay abreast of news about AIDS in greater depth than the popular press can provide. This article reviews current knowledge about the epidemiology, virology, and some clinical manifestations of AIDS.

Donald I. Abrams is Assistant Director, AIDS Activities Division, San Francisco General Hospital and Assistant Clinical Professor, Cancer Research Institute, University of California, San Francisco. Requests for reprints should be sent to: Donald I. Abrams, MD, San Francisco General Hospital, 995 Potrero Avenue, Ward 84, San Francisco, CA 94110. Donald I. Abrams is recipient of an American Cancer Society Development Award. Thanks to Karen Heller for her editorial assistance in preparing this manuscript.

VIROLOGY

AIDS encompasses the most severe manifestations of infection with the human immunodeficiency virus (HIV).[1,2] HIV is a retrovirus transmitted through infected blood and blood products, through shared unsterilized hypodermic syringes, and sexually through the exchange of infected genital secretions.[3,4]

The AIDS retrovirus has gone by several names since its discovery by investigators in France and the United States. These include human T-lymphotropic virus type 3, or HTLV-III; lymphadeno-pathy-associated virus, or LAV; and AIDS-associated retrovirus, or ARV.[5,6,7] In 1986, an international committee of viral taxonomists and AIDS investigators reached consensus on a name that more accurately describes the virus, the human immunodeficiency virus (HIV).[8]

The best current theory about the origins of HIV is that it derives from a simian or other animal virus that somehow came into the human population in Central Africa in the early 1970s. It is conceivable that with the rapid modernization and urbanization of the nations in this region, the virus spread from an isolated human population to urban centers where its expression became amplified. Support for this theory comes from research on the African green monkey species, which found that 40% carry a retrovirus closely related to HIV.[9] No other primate species examined shows evidence of a similar viral infection. Further evidence comes from the French discovery in West Africa of another human retrovirus, which appears to be a hybrid between the monkey virus and HIV.[10]

As a retrovirus that targets and ultimately destroys T-lymphocytes, the orchestra leaders of the immune system, HIV is particularly insidious.[11,12] A virus is the most basic form of life, a piece of genetic material, either DNA or RNA, surrounded by a protein coat to protect it from the environment. Retroviruses consist of a single piece of the intermediate messenger RNA. Genetic information normally flows from DNA to RNA; with retroviruses, the process is reversed. A unique enzyme, reverse transcriptase, allows the retroviral RNA to copy itself into DNA, which is then inserted into the genetic material of the infected cell where it remains for the life of the cell.

A foreign protein, or antigen, entering the body is initially processed by the so-called scavenger cells, or monocyte macrophages, which present the substance to either the B-cells or the T-cells, the lymphocyte subsets that form the basis of our immune system. B-cells respond by producing antibodies; T-cells neutralize the foreign invaders directly. T-cells are in two classes: the T-helper cell and the T-suppressor cell. Helper T-cells boost and augment the functions of the immune system; suppressor T-cells turn off the specific immune response. The helper T-cell provides positive feedback for the function of all the other cells in the system. Normally, healthy people have twice as many T-helper as T-suppressor cells; in people infected with HIV, however, helper T-cells are markedly depleted, and the normal 2:1 ratio of helper to suppressor cells is reversed.[13] With the loss of the helper T-cells, the body is subject to opportunistic infections caused by organisms to which we all have been exposed by which do not usually cause serious disease unless the immune system is impaired.

Other cellular targets of HIV include a subset of cells in the central nervous system.[14,15,16,17] Infection of these cells causes serious CNS dysfunction, including cerebral atrophy and many neuropsychiatric complications.[18,19]

EPIDEMIOLOGY

As of February, 1988, the Centers for Disease Control had recorded over 50,000 AIDS cases and almost 25,000 AIDS deaths nationwide. Perhaps as many as 1.5 to 2 million people in the United States are infected with HIV yet show no evidence of infection other than the presence of antibodies in their blood. Intermediate in number and degree of disease between healthy seropositives and patients with the life-threatening manifestations of AIDS is a large group of patients with the syndromes called AIDS-related complex, or ARC.[20] ARC is a term that was poorly defined and has been frequently misused. Many investigators hope it will be abandoned now that the Centers for Disease Control (CDC) have developed a more precise classification system for HIV infection.[1]

Groups at High Risk

When AIDS was first described in 1981, 95% of cases were diagnosed among homosexual or bisexual males;[21] that percentage has since fallen to about 74% of all AIDS cases.[22] Intravenous drug abusers (IVDAs), the second group recognized to be at particular risk for AIDS, account for 17% of cases.[23] Recently, the CDC began to differentiate in its reporting between homosexual and bisexual men who are not intravenous drug abusers (66%) and those who also have this additional risk factor (8%). Combining gay and heterosexual intravenous drug abusers, 25% of AIDS patients have a history of IVDA. With evidence of increasing HIV antibody seroprevalence rates among IVDAs in some cities, the proportion of AIDS cases in this risk group may be expected to increase in future years.[24,25,26,27] Retrospective studies suggest that once HIV is introduced into the IVDA population, it spreads rapidly. In New York City, for example, seroprevalence rates among IVDAs rose from 11% in 1977 to 27% in 1979 to 58% in 1984.[28] Similar rates of increase among IVDAs have been reported in European cities.[29,30,31,32]

Overall, young white males are disproportionately represented among AIDS cases, but among heterosexual IVDAs with AIDS, more than two-thirds are Black and Latino.[33,34] Of children with AIDS, 56% are Black and 24% are Latino. The vast majority of pediatric AIDS cases result from perinatal transmission, usually from infected mothers who are themselves IVDAs or the sexual partners of IVDAs.[35] In San Francisco, where 10% of intravenous drug abusers were HIV antibody positive in 1984 and 1985, seroprevalence was higher among Blacks (15%) and Latinos (14%) then whites (6%).[34] Racial differences in seroprevalence also have been reported in New York and New Jersey.[12]

Heterosexual Transmission

HIV can be transmitted sexually from males to females and females to males.[36,37,38,39] Although in the United States, 92% of AIDS cases has occurred in men and 8% in women, in Central Africa, where AIDS has had a devastating impact on heterosexuals, the sexual ratio of AIDS cases is 1.1 male to 1.0 female.[40,41,42] In the

United States, only 4% of AIDS cases has resulted from heterosexual transmission to date, but by 1991 this percentage is expected to increase to about 8%.[43]

Geographic Distribution

Since 1981, New York City has accounted for one-third of all AIDS cases reported in the United States, and San Francisco has had the second largest number of cases. With more AIDS diagnoses now reported in other parts of the country, San Francisco and New York each contribute slightly smaller proportions of cases to the total, a trend which is expected to continue. Compared to early in the epidemic when the number of new AIDS cases in San Francisco doubled every six months, then every nine months, the number of new cases has now slowed. Even so, in December 1986 three people per day (110/month) were diagnosed with AIDS in San Francisco and the same number per day (100/month) died of the disease.[44]

Relationship to Hepatitis B Studies

When AIDS was first recognized, the Centers for Disease Control became interested in the stored blood serum of 6,000 homosexual men who had participated in a hepatitis-B vaccine trial in San Francisco between 1978 and 1980. These men were highly sexually active and received their health care from a San Francisco venereal disease clinic. Once the HIV antibody test became available, the CDC in collaboration with the VD clinic began testing the stored blood samples from this cohort and found that in 1978, 4% of these men were infected with HIV.[45] By 1981, when we diagnosed our first AIDS cases in San Francisco, one-third of the "hepatitis-B" cohort was HIV antibody positive; by 1983, one-half of the cohort had seroconverted; and currently, 70% have seroconverted. From 1985-1986, however, the rate of seroconversion fell to zero. Interpreted optimistically, this may reflect the impact of the intensive educational efforts initiated in 1983 on changing high risk behaviors in the gay community. Or, it might simply indicate saturation of that population with the virus.

Projections for the Future

In June 1986, at a conference in West Virginia, representatives of the U.S. Public Health Service, the CDC, and other AIDS agencies and institutions predicted what the AIDS epidemic would look like five years hence.[43] The statistics that emerged are disturbing: by 1991, the cumulative total number of AIDS cases in the United States was estimated to increase tenfold. Calculated on the basis of 28,000 cases in June 1986, 270,000 cases were predicted by 1991. In that year alone, it was estimated that 54,000 Americans would die of AIDS, more than the total number of people killed during 11 years of the Vietnam War and roughly equal to the number of highway deaths nationwide each year.

CLINICAL MANIFESTATIONS
OF HIV INFECTION

Classification of HIV Infections

The Centers for Disease Control recently reclassified the diagnostic categories of HIV infection, primarily as a means of reclassifying patients with severe ARC into a disease category equivalent to that of patients with the opportunistic infections and malignancies characteristic of AIDS itself.[1] In the new schema, Group I consists of patients with the acute retroviral infection; Group II includes patients who are HIV antibody positive but have no other evidence of the retroviral infection; Group III consists of patients who have only persistent generalized lymphadenopathy (PGL). Group IV has several subcategories: IV-A includes patients with severe, disabling, sometimes fatal constitutional disease, such as those with the "wasting syndrome"; IV-B includes patients with neurologic symptoms as their only manifestation of HIV infection, either AIDS Dementia Complex, or a spinal cord process; IV-C includes patients with the AIDS opportunistic infections; and IV-D includes patients with malignancies, particularly Kaposi's sarcoma and non-Hodgkin's lymphoma. The CDC will soon expand its surveillance definition of AIDS to include Groups IV-A and IV-B.

Clinical Course

Within a few weeks following infection with HIV, an unknown percentage of infected people experience an acute flu-like illness. Our knowledge about acute HIV infection comes from Australia, where a group of homosexual men at risk for HIV infection were tested at six-month intervals for evidence of antibodies.[46] Of the men who seroconverted during the first six months of the study, 12 reported experiencing an acute infection that lasted for a mean of 8 days and was characterized by fevers, sweats, myalgias, arthralgias, nonspecific flu-like symptoms, malaise, and lethargy. A sore throat also was commonly part of the initial presentation. Other nonspecific findings included anorexia, nausea, vomiting, headaches, photophobia. Lymphadenopathy occurred in 75% of these patients. An erythematous macular rash that disappeared within four days was reported by 50% of patients.

Persistent generalized lymphadenopathy (PGL) is a common manifestation of HIV infection.[47,48] In San Francisco we became aware of this syndrome in 1979, when a large number of otherwise healthy gay men were referred to our hematology clinic with diffuse lymphadenopathy, splenomegaly, and constitutional symptoms suggesting possible lymphoma. Only benign reactive changes were found on lymph node biopsy. Patients were advised to modify their lifestyles to diminish stimulation of the immune system. In 1981, when we first saw people with AIDS, we noticed on physical examination of these patients that many also had generalized lymphadenopathy. Patients reported that they had had enlarged lymph nodes for about two years. We began a prospective natural history study of patients with what was then known as the "Gay Lymph Node Syndrome" and is now called persistent generalized lymphadenopathy (PGL). During a conference call in June 1983, an ad hoc working group of the National Institutes of Health attempted to further define the lymphadenopathy syndrome as one of several AIDS-related clinical features which came to be called AIDS-related Complex (ARC).[20] For the first three years of the natural history study, we believed that lymphadenopathy and AIDS were alternative clinical responses, most likely to a viral infection. Once the causative agent was identified in April 1984, we began to appreciate that healthy seropositives and people with ARC and AIDS were manifesting stages in a continuum of response to infection with HIV.

In terms of progression along the gradient of HIV infection, a proportion of patients get the acute syndrome, but a majority are only later identified as healthy seropositives. In those patients with the acute syndrome, 75% develop lymph node enlargement as part of the initial response to infection; of these 75%, some will continue with persistent generalized lymphadenopathy. From both the healthy seropositive and the PGL subsets, patients may go on to develop symptomatic disease. At the end of 36 months of follow up of our PGL cohort at San Francisco General Hospital, only 10% had progressed to AIDS;[49] now at the end of 5 years of follow up, a significant further number of patients have progressed to AIDS.[50]

Prognostic Indicators

In association with lymphadenopathy, prognostic indicators for progression to AIDS include oral *Candida* (thrush), and oral hairy leukoplakia. Hairy leukoplakia, a white wart-like lesion with small velvety, hair-like projections on its surface, is caused either by the Epstein-Barr virus, which also is responsible for mononucleosis and Burkitt's lymphoma, or the human papilloma virus, which causes condylomata.[51] The presence of oral thrush or hairy leukoplakia increases the risk of developing AIDs in the near future by a factor of 54.[52]

Opportunistic Infections and Malignancies

Numerous opportunistic infections are associated with AIDS, including parasitic infections (e.g., *Pneumocystis carinii* pneumonia, cryptosporidiosis, toxoplasmosis), bacterial infections (e.g., *Mycobacterium avium-intracellulare*), viral infections (e.g., invasive cytomegalovirus, invasive herpes simplex virus), and fungal infections (e.g., cryptococcal disease, histoplasmosis).[53] Malignancies associated with AIDS include Kaposi's sarcoma, central nervous system lymphoma, and high grade B-cell lymphoma.[54]

Pneumocystis carinii pneumonia (PCP) is the most common opportunistic infection, seen in 55% to 60% of AIDS patients, and by far the most common initial presenting diagnosis in AIDS. Kaposi's sarcoma (KS) is the most common malignancy. In New York and San Francisco, where the AIDS clinical syndrome is quite familiar to physicians, an HIV antibody test is rarely done on a patient with PCP or KS because those diagnoses are recognized to be definitive of AIDS. Elsewhere in the country, however, HIV antibody tests are used more frequently in supporting an AIDS diagnosis, although they may be somewhat superfluous.

Kaposi's sarcoma was first described over 100 years ago by a Hungarian dermatologist who noted purple lesions on the lower extremities of elderly men. In AIDS patients, KS has been found primarily among gay men. Why this malignancy is rarely if ever diagnosed among members of the other risk groups remains unclear. The use of inhaled nitrates (amyl nitrate and butyl nitrate) may be a co-factor in the development of AIDS-related KS among gay males.[58,59] The use of inhaled nitrates by gay men who have been seen at our clinic at San Francisco General Hospital has markedly declined in recent years. We also have seen a decline in the numbers of patients presenting with KS, although any relationship between that and decreased use of inhaled nitrates is as yet unknown.

Some patients referred to our clinic have mucosal KS, but no cutaneous lesions. Often, however, KS is quite disfiguring. Because the malignant cell involved is in the lining of the lymphatic vessels or the lymphatic endothelium, edema often results.[60] Elderly men with KS often developed swelling of the lower extremities before they got the violaceous lesions. AIDS patients with KS frequently have marked edema of the face, which is especially disturbing when the periorbital swelling prevents patients from opening their eyes.

The majority of our patients at San Francisco General Hospital do not die of Kaposi's sarcoma per se; rather, they usually develop and subsequently die of PCP or another opportunistic infection. KS is only a cause of morbidity and mortality when it involves the lungs.[61,62] The life span of patients with pulmonary KS is one and a half to two months following diagnosis. A subset of KS patients die neither from an opportunistic infection nor from pulmonary KS, but from the wasting syndrome (in Africa called the "slim" disease), which resembles an opportunistic infection in its symptomatology—fevers, night sweats, weight loss, profound diarrhea.[63] Patients with the wasting syndrome frequently develop AIDS Dementia Complex, which is caused by HIV infection of the brain and central nervous system.[45] Other clinical syndromes observed in people with AIDS and ARC which cannot yet be attributed to any opportunistic infection also may be directly related to HIV infection.

TREATMENT OF HIV INFECTION

Several immunomodulators and anti-viral drugs are being evaluated as possible treatments for HIV infection.[65,66,67,68] Because HIV preferentially infects activated or stimulated lymphocytes, immunomodulators potentially may worsen rather than alleviate immune system damage by stimulating infected T-cells. More promising therapies may consist of anti-viral drugs that neutralize reverse transcriptase.

One drug of this type, azidothymidine (AZT), recently was approved by the Food and Drug Administration for prescription to patients who meet certain diagnostic criteria. During clinical trials, AZT improved life quality and extended survival of patients who received the drug.[69,70] However, severe side effects, including bone marrow suppression in some patients, may limit its utility over the long term.[71]

Because people with AIDS already are at the end stages of viral infection, they are perhaps less appropriate for testing antiviral drugs than those with ARC or healthy seropositives. At the University of California, San Francisco, healthy seropositive homosexual men from the "hepatitis-B" cohort are being offered the opportunity to participate in a large clinical trial of AZT and acyclovir, a drug helpful in treating herpes virus infections. The trial is designed to determine whether early intervention with potentially active antivirals will slow or stop the progression of HIV-infected persons to AIDS. If these kinds of early interventions prove to be effective, they may offer compelling reasons to expand HIV antibody testing in order to identify people in the early stages of infection who might benefit from treatment. As yet, however, expanded antibody testing raises difficult ethical issues that weigh potential public health benefits of testing against potential harm to individual civil liberties.[72]

Moreover, because some fatalities have occurred in direct response to therapeutic interventions during clinical trials of drugs given to patients who only had ARC,[73] many ethical and clinical problems must be resolved before moving to large scale drug testing in healthier patients.

OUTCOME

Based on findings from our natural history study of patients with PGL and the experience of the San Francisco "hepatitis-B" cohort, the prognosis for people infected with HIV appears to be poor. Researchers at the San Francisco Department of Public Health have evaluated 60 men in the hepatitis-B cohort who have been antibody-positive since 1978. Of those infected with HIV for longer than 8 years, 60% now have ARC or AIDS.[64] The ultimate proportion that will advance to AIDS is unknown, since we have only been following this group prospectively for 5 years. The clinical trajectory of HIV infection observed thus can be likened to some extent to primary and secondary syphilis; what we have seen so far of AIDS may be the early manifestations of HIV infection. It is possible that in patients who survive 10 years or longer with HIV infection, the later disease will again be the equivalent of tertiary syphilis insofar as it may be primarily a central nervous system process.

The case fatality ratio for AIDS remains 50%. This means that at any given time, half the people who have been diagnosed with AIDS are dead. The average life span of an AIDS patient is one year from diagnosis: therefore, at any given point in time, roughly half the patients with AIDS have been diagnosed within the previous six to twelve months. Ultimately, however, the mortality rate of AIDS approaches 100%.

Those patients who present with an opportunistic infection alone have a higher case fatality rate than those who present initially with a malignancy. In San Francisco, the average age of our patients with Kaposi's sarcoma is 30 years, and their average life span following diagnosis is one and one-half years. The average age of our *Pneumocystis* patients is 40 years, and the average life span following diagnosis is 10 months.[55] The life span average has remained unchanged since the beginning of the epidemic.[56,57]

CONCLUSION

In the six years since AIDS was first recognized to be a new disease, we have learned a great deal about it. Although we still lack a vaccine or cure for the underlying HIV infection, we know enough about how HIV is and is not transmitted to educate people to protect themselves and others from exposure to the virus. In particular, people must be educated about safer sexual practices and the use of condoms, an effective barrier against HIV and other sexually-transmitted diseases.[74] Intravenous drug users should be advised to avoid sharing needles, but in the absence of readily available sterile syringes, they should be taught how to disinfect their syringes between each use.

Effective AIDS education and prevention are often hampered by a constellation of cultural taboos related to fears of human sexuality, disabling illness, mental illness, the alien "other," and death itself (D. Schulman, personal communication). Nevertheless, education remains our best defense against both misplaced fears and further viral spread.

NOTES

1. Centers for Disease Control. Classification system for human T-lymphotropic virus type III lymphadenopathy-associated virus infection. Morbidity Mortality Weekly Report 1986; 35:344-339.

Fauci AS. The acquired immunodeficiency syndrome: An update, Ann. Intern Med. 1985; 102:800-813.

3. Castro KG, Hardy AM, Curran JW. The acquired immunodeficiency syndrome: Epidemiology and risk factors for transmission. Med Clin North Am 1986; 70:635-649.

4. Goedert JJ and Blattner WA. The epidemiology of AIDS and related conditions. In: DeVita VT, ed. AIDS: Etiology, Diagnosis, Treatment and Prevention. Philadelphia: Lippincott, 1985:1-30.

5. Barre-Sinoussi F, Chermann J-C, Rey F et al. Isolation of a T-lymphotropic retrovirus from a patient at risk for acquired immunodeficiency syndrome (AIDS). Science 1983; 220:868-871.

6. Gallo RC, Salahuddin SZ, Popovic M et al. Frequent detection and isolation of cytopathic retroviruses (HTLV-III) from patients with AIDS and at risk for AIDS. Science 1984; 224:500-503.

7. Levy JA, Hoffman AD, Kramer SM et al. Isolation of lymphocytotropic retroviruses from San Francisco patients with AIDS. Science 1984; 225:840-842.

8. Coffin J, Haas A, Levy J et al. What to call the AIDS virus? Nature 1986; 321:10.

9. Barin F, M'Boup S, Denis F et al. Serological evidence for virus related to simian T-lymphotropic retrovirus III in residents of West Africa. Lancet 1985; 2:1387-1389.

10. Clavel F, Guetard D,' Brun-Vezinet F, Chamaret S, Rey M-A, Santos-Ferreira MD, Laurent AG, Dauguet C, Katlama C, Rouzioux C, Klatzmann D, Champalinmaud JL, Montagnier L. Isolation of a new human retrovirus from West African patients with AIDS. Science 1986; 233:343-346.

11. Dagleish AG, Beverly PCL, Clapham PR et al. The CD4 (T4) antigen is an essential component of the receptor for the AIDS retrovirus. Nature 1984; 312:763-767.

12. Klatzmann D, Champagne E, Chamaret S, et al. T-lymphocyte T4 molecule behaves as the receptor for human retrovirus LAV. Nature 1984: 312:767-768.

13. Ammann A, Abrams D, Conant M, Chudwin P, Cowan M, Volberding P, Lewis B and Casavant C. Acquired immune dysfunction in homosexual men: Immunologic profiles. Clin Immunol. Immunopathol 1983; 27:315-325.

14. Epstein LG, Sharer LR, Cho E-S, Myerhofer M, Navia BA, and Price RW. HTLV-III/LAV-like retrovirus particles in the brains of patients with AIDS encephalopathy. AIDS Res 1986; 1:447-454.

15. Koenig S, Gendelman HE, Orenstein JM, DelCanto MC, Pezeshkbour GH, Yungbluth M, Janotta F, Aksamit A, Martin MA, Fauci AS. Detection of AIDS virus in macrophages in brain tissue from AIDS patients with encephalopathy. Science 1986; 233:1089-1093.

16. Stoler MH, Eskin TA, Benn S, Angerer RC and Angerer LM. Human T-cell lymphotropic virus type III infection of the central nervous system: A preliminary in situ analysis. JAMA 1986, 256:2360-2364.

17. Gartner S, Markovits P, Markowitz DM, Betts RF and Popovic M. Virus isolation from and identification of HTLV-III/LAV-producing cells in brain tissue from a patient with AIDS. JAMA 1986; 256:2365-2371.

18. Levy RM, Bredesen DE, Rosenblum ML. Neurological manifestations of the acquired immunodeficiency syndrome (AIDS): Experience at UCSF and review of the literature. J. Neurosurg 1985; 62:475-495.

19. Navia BA, Cho ES, Petit CK, Price RW. The AIDS dementia complex. Ann. Neurol 1986; 19:525-535.

20. Abrams, DI. AIDS-related conditions. In: Pinching A, ed. Clinics in Immunology and Allergy. London: WB Saunders, 1986:581-599.

21. Centers for Disease Control. Kaposi's sarcoma and Pneumocystis pneumonia among homosexual men—New York City and California. Morbidity Mortality Weekly Report 1981; 30:305-308.

22. Centers for Disease Control. Update on Kaposi's sarcoma and opportunistic infections in previously healthy persons—United States. Morbidity Mortality Weekly Report 1982; 11:294-301.

23. Centers for Disease Control. Update: acquired immunodeficiency syndrome — United States. Morbidity Mortality Weekly Report 1986; 35:17-21.

24. Friedland GH, Harris C, Butkus-Small C, Shine D, Moll B, Darrow W, Klien R. Intravenous drug abusers and the acquired immunodeficiency syndrome (AIDS): Demographic, drug use and needle-sharing patterns. Arch Intern Med 1985; 8:1413-1437.

25. San Francisco Department of Public Health. AIDS in IV drug users, San Francisco, 1979-1986. San Francisco Epidemiology Bult 1986; 2(5):1-2.

26. Chaisson RE, Moss AR, Onishi R, Osmond D, Carlson JR. Human immunodeficiency virus infection in heterosexual intravenous drug users in San Francisco. Am J Public Health 1987; 77:169-172.

27. Marmor M. Des Jarlais DC, Cohen H, Friedman SR, Beatrice ST, Dubin N, El-Sadr W, Mildvan D, Yancovitz S, Mather U and Holtzman R. Risk factors for infection with human immunodeficiency virus among intravenous drug abusers in New York City. AIDS 1987; 1:39-44.

28. Weiss SH, Ginzberg HM, Goedert JJ. Risk of HTLV-III exposure and AIDS among parenteral drug abusers in New Jersey. Proceedings First International Conference on AIDS, Atlanta 1985; abstract.

29. Angarano G, Pastore G, Monno L, Santantonio J, Luchena N, Schiraldi O. Rapid spread of HTLV-III infection among drug addicts in Italy. Lancet 1985; 2:1302.

30. Robertson JR, Buckwall AB, Welsby PD et al. Epidemic of AIDS related virus (HTLV-III/LAV) infection among intravenous drug abusers. Br. Med J 1986; 292:527-529.

31. Tirelli U, Vaccher E, Carbone A et al. Heterosexual contact is not the predominant mode of HTLV-III transmission among intravenous drug abusers. JAMA 1986; 255:2289.

32. Rodrigo JM, Serra MA, Aguilar E, DelOlmo JA, Gimeno V, Apasisi L. HTLV-III antibodies among drug addicts in Spain. Lancet 1985; 2:156-157.

33. Bakeman R, Lumb JR. AIDS risk group profiles in whites and members of minority groups. NEJM 1986; 314:191-192.

34. Peterson JL and Andrews E. AIDS and blacks: Gay identity, racial poverty and racial discrimination. Multi-cultural Inquiry and Research on AIDS Newsletter 1987; 1:3-4.

35. Scott GB, Fischl MA, Klimas SN, Fletcher MA, Dickinson GM, Levine RS, Parks WP. Mothers of infants with the acquired immunodeficiency syndrome: Evidence for both symptomatic and asymptomatic carriers. JAMA 1985; 253:363-366.

36. Harris C, Butkus-Small C, Klein RS. Friedland GH, Moll B, Emerson EE, Spigland I and Steigbigel, NH. Immunodeficiency in female sexual partners of men with the acquired immunodeficiency syndrome. 1983; 308:1181-1184.

37. Des Jarlais DC, Chamberland ME, Yancovitz SR et al. Heterosexual partners: a large risk group for AIDS. Lancet 1984; 2:1346-1347.

38. Centers for Disease Control. Heterosexual transmission of human T-lym-

photrophic virus type III/lymphadenopathy-associated virus. Morbidity Mortality Weekly Report 1985; 34:561-563.

39. Calabrese LH, Gopalakrishna KV. Transmission of HTLV-III infection from man to woman to man. NEJM 1986; 314:987.

40. Clumeck N, Sonnet J, Taelman H et al. Acquired immunodeficiency syndrome in African patients. NEJM 1984; 310:492-97.

41. Piot P, Quinn TC, Taelman H, Feinsod FM, Minlangy KB, Wob MO, Mberdi N, Mazebo P, Ndangi K, Stevens W, Kalambayi K, Mitchell S, Bridts C, McCormick JB. Acquired immunodeficiency syndrome in a heterosexual population in Zaire. Lancet 1984; 2:65-69.

42. Clumeck N, Robert-Guroff M, Van de Perre P, Jennings A, Sibomanz J, Sibomana P, Cran S, Gallo RC. Seroepidemiological studies of HTLV-III antibody prevalence among selected groups of heterosexual Africans. JAMA 1985; 254:2599-2602.

43. Coolfont Report: A PHS plan for prevention and control of AIDS and the AIDS virus. Pub Health Rep 1986; 101:341-348.

44. San Francisco Department of Public Health. Current status and projections. San Francisco Epidemiol Bulletin 1986; 2(11):1-4.

45. Jaffe HW, Darrow WW, Echenberg DF, O'Malley PM, Getchel JP, Kalyanaraman VS, Byers RH, Drennan DP, Braff EH, Curran JW, Francis DP. The acquired immunodeficiency syndrome in a cohort of homosexual men: a six year follow-up study. Ann Intern Med. 1985; 103:210-214.

46. Cooper DA, Gold J, Maclean P et al. Acute AIDS retrovirus infection. Lancet 1985; 1:537-540.

47. Metroka CE, Cunningham-Rundles S, Pollack MS, et al. Persistent generalized lymphadenopathy in homosexual men. Ann Intern Med 1983; 99:585-591.

48. Abrams DI, Lewis BJ, Beckstead JH, Casavant CA, Drew WL. Persistent diffuse lymphadenopathy in homosexual men: Endpoint or prodrome? Ann Intern Med 1984; 100:801-808.

49. Abrams DI, Mess T and Volberding P. Lymphadenopathy: endpoint or prodrome? Update of a 36 month prospective study. In Gupta S (ed) *AIDS Associated Syndromes. Advances in Experimental Medicine and Biology*, Vol. 187 New York: Plenum Press, 1985: 73-84.

50. Abrams DI, Kirn, DH, Feigal DW and Volberding PA. Lymphadenopathy: Update of a 60 month prospective study. Proceedings III International Conference on AIDS, Washington DC 1987; 118 (abstract).

51. Greenspan JS, Greenspan D, Lennette ET, Abrams, DI, Conant MA, Petersen V, Freese UK. Epstein-Barr virus replicates within the epithelial cells of oral "hairy" leukoplakia, an AIDS-associated lesion. N. E. J. M. 1985; 313:1564-1571.

52. Greenspan D, Greenspan SS, Hearst NG, Pan L-Z, Conant MA, Abrams DI, Hollander H, Levy JA. Oral hairy leukoplakia: human immunodeficiency virus status and risk for development of AIDS. J Inf Dis 1987; 55:475-481.

53. Grant IH and Armstrong P. Management of infectious complications in acquired immunodeficiency syndrome. Am J Med 1986; 81 (suppl 1A):59-72.

54. Volberding PA. Kaposi's sarcoma, B-cell lymphoma and other AIDS-associated tumors. In: Pinching A, ed. Clinics in Immunology and Allergy. London: WB Saunders, 1986:569-580.

55. Moss AR, McCallum G, Volberding PA et al. Mortality associated with mode of presentation in the acquired immunodeficiency syndrome. JNCI 1984; 73:1281-1284.

56. Lemp GF, Barnhart JL, Rutherford GW and Werdeger D. Predictors of survival for AIDS cases in San Francisco. Proceedings III International Conference on AIDS. Washington DC 1987; 118 (abstract).

57. Volberding P, Feigal DW, Cutler K and Hearst N. Decreasing survival in recently diagnosed AIDS-related Kaposi's sarcoma. Proceedings III International Conference on AIDS, Washington DC. 1987; 172 (abstract).

58. Newell GR, Adams SC, Mansell PW et al. Toxicity, immunosuppressive effects and carcinogenic potential of volatile nitrites: Possible relationship to Kaposi's sarcoma. Pharmacotherapy 1984; 41:284-297.

59. Haverkos HW, Pinsky PF, Drotman DP et al. Disease manifestations among homosexual men with acquired immunodeficiency syndrome: A possible role of nitrites in Kaposi's sarcoma. J. Sex Trans Dis 1985; 12:203-208.

60. Beckstead JH, Wood GS, Fletcher V. Evidence for the origin of Kaposi's sarcoma from lymphatic endothelium. Am J Pathol 1985; 119:294-300.

61. Kaplan LD, Jaffe H, Hopewell D, Abrams DI, Volberding PA. Pulmonary Kaposi's sarcoma (PKS) in AIDS. Proc. Am. Soc. Clinc. Onc., 1985; 4:4 (abstract).

62. Medur GU, Stover DE, Lee M, Myskowski PL, Caravelli JF and Zaman MB. Pulmonary Kaposi's sarcoma in the acquired immune deficiency syndrome. Am J Med 1986; 81:11-18.

63. Serwadda D, Mugerwa RD, Sewankambo NK et al. Slim disease: a new disease in Uganda and its association with HTLV-III infection. Lancet 1985; 2:1849-1852.

64. Hessol NA, Rutherford GW, O'Malley PM, Doll LS, Darrow WW, Jaffe HW. The natural history of human immunodeficiency virus infection in a cohort of homosexual and bisexual men: a 7 year prospective study. Proceedings III International Conference on AIDS. Washington DC, 1987; 1 (abstract).

65. Kaplan LD, Wofsy CB, Volberding PA. Treatment of patients with acquired immunodeficiency syndrome and associated manifestations. JAMA 1987, 257:1367-1374.

66. Fauci AS (moderator). Immunomodulators in clinical medicine. Ann Intern Med 1987; 106:421-433.

67. DeVita VT Jr. (moderator). Developmental therapeutics and the acquired immunodeficiency syndrome. Ann Intern Med 1987; 106:568-581.

68. Yarchoan R, Broder S. Strategies for the pharmacologic intervention against HTLV-III/LAV. In: Broder S. ed. AIDS: Modern concepts and therapeutic challenges. New York: Marcel Dekker 1986:335-360.

69. Yarchoan R, Klecker RW, Weinhold KJ et al. Administration of 3'-azido-

d'-deoxythymidine, an inhibitor of HTLV-III/LAV replication to patients with AIDS or AIDS-related complex. Lancet 1986; 1:575-580.

70. The AZT Collaborative Working Group. The efficacy of azidothymidine in the treatment of patients with AIDS and AIDS-related complex: a double-blind placebo-controlled trial. Proceedings III International Conference on AIDS. Washington DC. 1982; 101 (abstract).

71. The AZT Collaborative Working Group. The toxicity of 3'-azido-3'deoxythymidine (azidothymidine) in the treatment of patients with AIDS and AIDS-related complex: a double-blind placebo-controlled trial. Proceedings III International Conference on AIDS. Washington DC 1987; 58 (abstract).

72. Novick A, Dubler NN, Landesman SH. Do research subjects have the right to know their HIV antibody test results? IRB: A review of Human Subjects Research 1986; 8:6-9.

73. Kaplan LD, Wolfe PR, Volberding PA, Feorino P, Levy JA, Abrams DI, Kiprov D, Wong R, Kaufman L, Gottlieb MS. Lack of response to suramin in patients with AIDS and AIDS-related complex. Am J Med., 82:615-620.

74. Conant M, Hardy D, Sernatinger J, Spicer D, Levy JA. Condoms prevent passage of AIDS-associated retrovirus. JAMA 1986; 255:1708.

Alcohol and Drugs
as Co-Factors for AIDS

Rob Roy MacGregor, MD

I. INTRODUCTION

In discussing the potential effects of alcohol and drug use on the AIDS epidemic, it must be stated first that no direct evidence exists currently to indicate an association between alcohol use and the development of the Acquired Immunodeficiency Syndrome (AIDS). Moreover, the strong association between intravenous illicit drug use and AIDS can be ascribed to the direct inoculation of infected blood rather than to any indirect effects of the drugs on the host. Nonetheless, because of evidence that alcohol and other drugs can inhibit the immune system in various ways, it is reasonable to consider potential mechanisms through which these agents could pharmacologically modify the interaction of the Human Immunodeficiency Virus (HIV) with the human host. First, we will consider several stages of the host-virus interaction at which alcohol and drugs could act, and the potential impact that modulation at these stages might have. Then we will examine sequentially the various arms of immune defense and review the data concerning the effects of alcohol and other drugs on their function.

From the Infectious Diseases Section, Department of Medicine, University of Pennsylvania School of Medicine, Philadelphia, PA 19104.

Supported in part by Grant #AA-06029, National Institute on Alcohol Abuse and Alcoholism.

II. STAGES OF RISK

Alcohol and other drugs could be physiological risk factors for the development of AIDS in two different ways: A. they could increase the risk of primary infection when the individual is first exposed to the Human Immunodeficiency Virus (HIV); B. for individuals already infected with HIV, they could depress the immune mechanisms which act to limit its negative impact, resulting in a progression from asymptomatic to clinical infection.

A. Infectivity

Defense against virus infection is primarily through humoral and cell-mediated immune mechanisms.[1] Therefore, it is possible that an individual with one or more of these arms of immunity temporarily suppressed by drugs might not be able to prevent the virus from successfully achieving a primary infection. However, the contrary argument also could be made: HIV uses the T4 antigen as a ligand, and, as a result, infects cells which express it.[2] Thus, theoretically, any agent which depresses cell-mediated immunity either by decreasing the number of circulating T-4 cells or by decreasing the expression of T4 antigen on lymphocytes, would reduce the number of binding sites for the HIV, and therefore could diminish the risk of infection. This thesis is completely speculative, and is raised solely to underscore the distressing lack of *any* information regarding the effect of alcohol or other drugs on the resistance of animals to primary viral infection. Other than by direct injection of HIV through needle sharing, the only way in which alcohol and/or drug use is known to potentiate the risk of developing an HIV infection is through their disinhibiting effect. Individuals often take these drugs in order to lose their inhibitions. The result is an increased willingness to participate in "risky behaviors" which increase the chances for their exposure to the HIV.

B. Progression to Symptomatic Disease

Drugs might also affect the interaction of HIV with the human host *after* the infection has been established. Most of the people currently infected with HIV are totally asymptomatic, and may be identified only through a serum antibody test. The 1987 estimate for this population's size in the U.S. is approximately 2 million people. For about 20% of individuals within this large group, the infection has progressed to the point where they have some symptoms: weight loss, fever, diarrhea, lymph node enlargement, etc. (Figure 1). At this stage, the infection causes a symptom complex which has been called Progressive Generalized Lymphadenopathy with Wasting, or AIDS-Related Complex (ARC). The precise character-ization of the syndrome is less important than the recognition that these individuals have not been able to control the infection at an asymptomatic stage, and that it has progressed to a clinical illness.

Figure 1

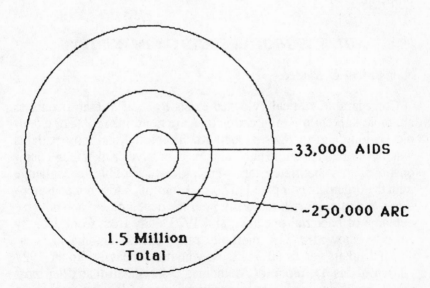

APRIL 1987

33,000 AIDS

~250,000 ARC

1.5 Million
Total

HIV-INFECTED POPULATION

Finally, within the symptomatic group is a yet smaller population of infected individuals, those whose HIV infection has caused sufficient damage to their cell-mediated immunity that they have developed an infection or neoplasm characteristic of patients with immunodeficiency; these patients have AIDS. One of the biggest mysteries of the epidemic to date is the identity of the factors which cause individuals to move from the large infected-but-asymptomatic pool into the group with symptoms; moreover, once symptomatic, what causes them to develop full-blown AIDS. The progression represents a sequential loss of T-cell function,[3] and appears to indicate a progressive failure of the host's immune system to control the HIV, to prevent its destructive spread to uninfected cells in the body. At present, no convincing epidemiologic data exist which link the use of alcohol or other drugs to a progression of the HIV infection from the asymptomatic stage to ARC or AIDS. However, it seems likely that illnesses and agents which do interfere with normal immune defenses might have such an effect. In the following sections, we shall review the evidence that alcohol and other drugs should be considered as agents with immunosuppressive properties.

III. ALCOHOL EFFECTS ON IMMUNITY

A. Clinical Evidence

Considerable clinical evidence exists to suggest that alcoholics are more susceptible to infection than are nondrinkers.[4-6] For example, as long ago as 1884, Robert Koch reported that most patients who developed epidemic cholera were excessive drinkers, and demonstrated that inebriated rats were more susceptible to challenge with the organism compared to normal controls.[7] Alcohol consumption is a particular risk factor for both the development and adverse outcome of bacterial pneumonia: a 1923 study from Cook County Hospital showed a 22% mortality rate from pneumonia in occasional drinkers versus 50% in those who drank heavily.[8] In the 1927 edition of his Textbook of Medicine, Osler stated that "the most potent predisposing factor to pneumonia is the lowered resistance due to alcohol," and quoted his own figures of 18.5% mortality

from pneumonia among abstainers compared to 52.8% in alcoholics. In 1965, prospective analysis of 900 consecutive admissions
to Yale-New Haven Hospital indicated that 16.5% of alcoholics had
bacterial pneumonia, compared to a rate of 6.5% for the rest of the
hospital population.[9] More precise, population-based rates of pneumonia mortality were generated in a recent Canadian study: the ratio of observed to expected pneumonia deaths was 3.1:1 for alcoholic men and 7.1:1 for alcoholic women.[10] Alcoholism is also a
predisposing factor in 30% of patients with aspiration lung ab-
scess.[11] When alcoholism is complicated by cirrhosis of the liver,
the patients manifest a unique tendency to develop spontaneous
bacteremias and peritonitis, particularly with pneumococci and en-
teric gram-negative bacilli.[12] Finally, the association between alcoholism and tuberculosis is well documented. Population-based data
from the U.S., Scandinavia, and Australia all indicate that alcoholics have a much greater frequency and severity of tuberculosis
than that found in the general population.[6] Thus, numerous clinical
clues suggest that alcohol consumption may interfere with normal
host responses to infection. Next, we shall examine evidence for
specific immune dysfunction associated with alcohol use, organizing our review by the various components of host defense. The
reader may also wish to consult several excellent articles on the
subject.[4-6]

B. Cell-Mediated Immunity

The time-honored method of assessing cell-mediated immunity
(see Table 1) is the delayed hypersensitivity reaction to intrader-
mally-administered antigen. Comparison of skin test reactivity between alcoholics and normal controls has shown fewer and smaller
reactions to a range of antigens.[13-15] The degree of inhibition correlates with the extent of liver damage, which in turn is often associated with the degree of protein-calorie malnutrition.[13] Moreover,
reactivity increases on discontinuation of alcohol exposure, re-
sumption of good nutrition, and improvement of liver function.
Neither acute intoxication[16] nor chronic drinking in a protected
study environment (accompanied by good nutrition)[17,18] affect the
frequency or size of skin test reactions in patients who have previ-

Table 1

ALCOHOL EFFECTS ON CELL-MEDIATED IMMUNITY

	Lymphocytes	Macrophages/Monocytes	RES Clearance
In Vivo			
Acute:	Normal DH	↓ pulmonary macrophage mobilization	↓ peritoneal, hepatic, pulmonary
Chronic:	↓ DH with liver disease ↓ DH sensitization Lymphopenia ↓ T and NK cells with liver disease	Unknown	↓ hepatic
In Vitro:	↓ Lymphocyte transformation ↓ Lymphocyte migration ↓ NK cell activity ↓ ADCC	↓ alveolar macrophage adhesion, phagocytosis and bactericidal activity ↓ monocyte Fc receptors, phagocytosis	

ously demonstrated delayed hypersensitivity to a given antigen. Moreover, sensitization to new antigens was found to proceed normally in noncirrhotic patients admitted for alcohol withdrawal.[19] In contrast, nondrinking patients with cirrhosis[14,20] are inhibited in their ability to develop primary cell-mediated immunity. Moreover, noncirrhotic alcoholics maintaining good nutrition while drinking chronically on a Study Ward failed to become sensitized to a primary antigen;[17] similar blocking of primary immunization has been reported in rats taking alcohol for prolonged periods.[21] Thus, established delayed hypersensitivity is resistant to suppression by acute or chronic alcohol ingestion, whereas the establishment of new primary cellular-immune responsiveness is vulnerable. In addition, protein-calorie malnutrition, which is known to depress both sensitization and recall responses, may potentiate the effect of alcohol in some patients.

The total number of circulating lymphocytes is decreased in some alcoholics drinking in an uncontrolled environment, with return to normal levels within several days of abstinence.[22-24] However, controlled studies in humans and animals, in which good nutrition was maintained during drinking periods of up to 42 days, have failed to demonstrate lymphopenia in any participants. Measure of the actual number of thymus-dependent lymphocytes has only been carried out in alcoholics with concomitant liver disease.[20,25,26] A decrease in T-lymphocytes is found, correlating directly with the severity and acuteness of the hepatocellular injury; most pronounced in acute alcoholic hepatitis, and less so in conditions of stable, mild cirrhosis.

Alcohol has been shown to depress a number of in vitro lymphocyte function parameters. For example, blast transformation of normal lymphocytes to several mitogens is impaired when conducted in the presence of alcohol concentrations commonly found in drinkers.[27-29] Findings with cells from alcoholics are more complex: response to phytohemagglutinin, a nonspecific mitogen, is poor in patients with various degrees of chronic liver disease,[18,19] although acute alcoholic hepatitis is associated with normal or even increased responsiveness to transforming stimuli.[14,30] This enhancement of activity may relate to a loss of suppressor T-cell activity, reported in another study of alcoholic hepatitis.[31] To complicate the issue further, a serum factor which prevents mitogen-induced lymphocyte transformation has been found in patients with alcoholic cirrhosis of the liver, correlating with the degree of lymphopenia.[32,33] Another serum "factor" of potential major importance is alcohol itself; none of the studies utilizing cells or serum from alcoholic patients have included "clinical" concentrations of alcohol in the test systems. Because its inhibitory effects might be reversible on withdrawal of cells from the alcohol environment, the studies to date may be underestimating the frequency of alcohol-induced dysfunction.

Natural killer (NK) cytotoxic activity of normal lymphocytes is also inhibited by alcohol in vitro, perhaps secondary to reduced binding to target cells.[34,35] Cells from subjects who have been drinking chronically, tested in an alcohol-free environment, have shown normal or even increased NK cell activity,[36,37] although nondrinking cirrhotic patients have depressed activity, particularly if malnourished.[38] As with lymphocyte transformation, it would be most informative to test patients' cells in the presence of alcohol. Other lymphocyte functions which have been shown to be inhibited by in vitro exposure to alcohol include antibody-directed cellular cytotoxicity (ADCC),[39] and directed and random migration.[40] A potential mechanism which has been proposed for alcohol's inhibiting activity is its induction of increased intracellular cyclic AMP production in lymphocytes.[41]

In summary, alcohol has significant inhibitory activity against a range of cell-mediated immune functions. Alcoholic patients have decreased delayed hypersensitivity to skin test antigens, associated with reduced numbers of circulating T-cells and impaired lymphocyte transformation. During periods of prolonged drinking, development of immune responsiveness to new antigens is inhibited, although abstinent alcoholics undergo primary sensitization normally. Moreover, incubation of normal patients' lymphocytes with alcohol inhibits their transformation to mitogens, natural killer activity, and migration. Cirrhosis of the liver and malnutrition appear to be potentiating factors for alcohol-induced inhibition of cell-mediated immunity. Based on these facts, there is reason for concern that alcohol ingestion might reduce the effectiveness of an individual's immune defense against the HIV.

C. Pulmonary and Reticuloendothelial Clearance

As can be inferred from the high frequency of pneumonia in imbibers, alcohol interferes with pulmonary immune defense. Acutely intoxicated experimental animals clear aerosolized bacteria at a decreased rate compared to normal controls.[42] Mechanisms for this defect include depressed ciliary function,[43] inhibition of surfactant production,[44] poor migration of PMNs into the lung,[45] and depression of alveolar macrophage functions including mobilization of the macrophages into the lung,[46] and their in vitro adherence, phagocytosis, and bacterial killing.[47,48] Macrophages are vulnerable to inhibition by alcohol at other sites as well: intoxication slows bacterial clearance rates by peritoneal macrophages,[49] and liver[50,51] in rats, and one study in humans demonstrated that alcoholics admitted for detoxification had significantly slowed clearance rates for intravenously-administered aggregated albumin, which improved to normal within 7 days after withdrawal.[52] In addition, studies with human blood monocytes (circulating macrophages) have shown that incubation in vitro with alcohol depresses both their Fc receptor[53] and phagocytic[50] functions. Thus, alcohol exposure inhibits RES clearance, most likely as a result of inhibiting macrophage mobilization, activation, and phagocytosis.

D. Humoral Immunity

Antibody response to immunization with a new antigen is inhibited by concurrent alcohol consumption[17,21,54,55] (see Table 2). This was demonstrated first in 1909 by Parkinson, using chronically intoxicated rabbits,[54] and then confirmed in a rat model.[21,55] Humans drinking chronically in a controlled environment, with good nutrition, also have failed to develop antibody following primary exposure to a new antigen.[17] In contrast, recall anamnestic responses to antigens already experienced before the drinking period were not impaired.[21,55] Suppression of immunization requires concurrent alcohol ingestion; nondrinking alcoholics, with or without associated liver disease, have normal[55,56] or even increased[58] antibody responses to immunization with tetanus toxoid or pneumococcal polysaccharide.

Table 2

ALCOHOL EFFECTS ON HUMORAL IMMUNITY

	Immunoglobulins	Complement
In Vivo		
Acute:	Unknown	? ↓ bactericidal activity
		No change in CH-50
Chronic:	↓ primary antigen response	No change in CH-50
	Normal anamnestic response	
	Non-drinking cirrhotics have ↑ serum levels, ↑ spontaneous B-cell production, normal primary antigen response	

Serum concentrations of immunoglobulins are elevated in patients with alcoholic liver disease.[15,58-60] Because their gamma-globulin serum half-life is shortened, the most likely mechanism for the high concentration is an increase in production.[6] In support of this thesis, in vitro assessment of their lymphocytes indicates that, although the total numbers of B-cells are normal, they exhibit markedly increased spontaneous immunoglobulin production and reduced suppressor activity in vitro.[61,62] The stimulus for this enhanced production appears to be a persistent antigenic challenge as a result of portosystemic blood shunting that bypasses the liver's Kupffer cells, preventing them from removing antigen from circulation before it reaches the antibody-producing tissues.[63] Evidence to support this thesis is the increased antibody titers to *E. coli* antigens found in humans with cirrhosis, titers which rise still higher following surgical portocaval anastomosis.[64]

The effect of alcohol consumption on the complement system has not been examined adequately to date. Two controlled studies have shown that total hemolytic complement activity did not change in human volunteers undergoing acute intoxication[16] or chronic drinking in a controlled environment.[17] Uncontrolled studies have reported complement levels in alcoholics to be both reduced (especially in patients with cirrhosis)[65] and elevated.[59] Several groups have pointed to decreased bactericidal activity of serum following acute intoxication to suggest that alcohol interferes with complement activity.[66-68]

To summarize: Concomitant ingestion of alcohol blocks primary antibody responses to new antigens, whereas recall anamnestic responses occur normally. Total immunoglobulin production and serum concentrations are increased in patients with alcoholic liver disease, as a result of shunting of antigen past the liver. Complement data are too incomplete and conflicting to allow for any generalizations.

E. Polymorphonuclear Leukocyte Function

Although considerable evidence exists for the suppression of multiple facets of PMN function by alcohol, this arm of immune defense only plays a minor role against virus infection, and so is being considered last (See Table 3.) At present, there is no evidence to suggest that PMNs play any role in the host response to primary infection with HIV or in the control of its progression to ARC or AIDS. Moreover, with the exception of progressive neutropenia, it appears that HIV infection has no inhibiting effect on PMN function. Using the types of infections which develop as an indication of the hosts' immune competence, one is impressed with the infrequency of bacterial infections (controlled by PMNs) in AIDS patients, and with their overwhelming vulnerability to fungi, mycobacteria, herpes viruses, and parasites, organisms usually controlled by cell-mediated immunity. However, to make the review of alcohol effects on immunity complete, and because it has been shown that PMNs can exert antibody-directed cellular cytotoxicity against virally-infected cells, we shall briefly examine the effects of alcohol on PMN function, dividing the data into Production, Delivery, and Killing functions.

1. PMN Production

Mild levels of granulocytopenia, which correct over several days of abstinence, have been described in 4-8% of alcoholics who require hospitalization.[22,23] The mechanism appears to be marrow depression: histologic examination of marrow aspirates show a de-

Table 3

ALCOHOL EFFECTS ON NEUTROPHIL FUNCTION

	Production	Delivery	Phagocytosis/killing
In Vivo			
Acute:	No effect	↓ delivery ↓ adherence Normal chemotaxis	Normal
Chronic:	Neutropenia Marrow suppression	↓ chemotaxis ↓ delivery, adherence only with high blood levels	Normal
In Vitro:	↓ Marrow colonies ↓ CSF production	↓ adherence Normal chemotaxis	Normal

crease in mature PMNs, with vacuolization of myeloid precursors.[23,24] Although megaloblastic anemia and reduced serum folic acid levels frequently accompany the neutropenia,[23] alcohol itself has been shown to suppress in vitro granulocyte colony proliferation, even in the presence of excess folate.[27,69] This inhibition is a result of alcohol's suppression of colony-stimulating factor production by T-lymphocytes, rather than a direct alcohol effect on the myeloid line.[69]

2. PMN Delivery

Studies first performed in rabbits in 1938,[70] and later confirmed in man[71,72] have shown that acute intoxication blocks the delivery of PMNs to sites of local infection or sterile tissue trauma. This effect is time-limited, and relates directly to the blood alcohol concentration; chronic ingestion of subintoxicating doses of alcohol is not inhibitory.[17] The following evidence suggests that the mechanism for this delivery defect is reduced PMN adherence to endothelial cells, thereby preventing their diapedesis and movement into the extravascular space: (a) Incubation of whole blood in vitro with alcohol produces a dose-dependent inhibition of PMN adherence.[73] (b) In vivo alcohol administration also reduces adherence in a dose-dependent manner, correlating with the extent of inhibited local delivery.[72-74] (c) Correction of the adherence defect by pharmacologic means also corrects the delivery defect, and increases survival of animals challenged by bacterial infection while intoxicated.[74] Chemotaxis, the directed migration of PMNs toward a chemical signal, does not appear to be affected directly by alcohol. PMNs from acutely intoxicated normal volunteers demonstrate normal in vitro chemotaxis, and incubation of normal PMNs in vitro with intoxicating concentrations of alcohol also has no effect on chemotaxis.[16] However, both alcoholics admitted for withdrawal from chronic alcohol use,[59] and those drinking chronically on a study ward[17] have shown decreased in vitro chemotaxis. This probably results from an alcohol effect on the liver because patients drinking to the point of alcoholic hepatitis or cirrhosis frequently develop a serum inhibitor of the chemotactic factor C5a.[75-78] However, the in vivo significance of this inhibitor must be questioned in light of the recent report that patients with advanced cirrhosis and high circulating concentrations of the inhibitor had normal PMN mobilization into experimental skin abrasions in vivo.[79]

3. Killing

Once delivered to sites of inflammation, it appears that PMNs are able to phagocytize and kill bacteria in the presence of profoundly intoxicating alcohol concentrations.[9,17,59,70-72,80] Cells have been examined following exposure to alcohol in vivo by acute intoxication of normal volunteers[16,71] and from patients after periods of prolonged daily drinking;[17,59] in addition, phagocytosis and killing of PMNs from normal volunteers remain normal in alcohol concentrations up to 500mg%.[9,70,71,80] However, one study suggests that patients with alcoholic cirrhosis have a serum factor which inhibits post-phagocytic oxygen-dependent bactericidal events such as superoxide generation and bacterial iodination.[78]

In summary, alcohol inhibits PMN function at the points of production and delivery. Decreased colony-stimulating factor is associated with reduced marrow production and neutropenia, particularly in response to infection. Moreover, intoxicating concentrations of alcohol prevent normal PMN adherence to endothelial cells, thereby blocking local PMN delivery to sites of infection. Once on site, PMNs phagocytize and kill normally in the face of high concentrations of alcohol.

IV. EFFECTS OF OTHER DRUGS ON IMMUNITY

It is estimated that, currently, over three quarters of a million Americans are illicit users of parenteral drugs, and that 20% of their hospitalizations are for infectious complications of this drug use.[81] Therefore, it is reasonable to question the adequacy of their immune defenses. Although a large number of studies have been published in the last 20 years regarding the effects of different drugs on various immune functions, it remains difficult to state any general conclusions. On the one hand, the most carefully controlled studies have been done in experimental animals, particularly in rodents, but the wisdom of extrapolating these findings to man remains arguable. On the other hand, studies in man historically have been flawed by the possibility of concomitant use of other drugs, and, more recently, by the potential of coexistent asymptomatic HIV infection which could be responsible for any immunologic abnormalities found. With these problems in mind, let us now review and summarize available data regarding the effects of drugs on immune function, organizing our review by type of drug: opiates, marihuana, stimulants, barbiturates, and finally, amyl nitrite.

A. Opiates

One subacute exposure study in mice showed that five days' administration of high doses of morphine resulted in a decrease in spleen weight and cellularity, with the germinal centers appearing smaller and less distinct.[82] Primary antibody response was not affected by this short exposure, and, unfortunately, no tests of cell-mediated immunity were performed. Several groups of heroin/methadone addicts have been evaluated, but none of the reports include data on the subjects' HIV antibody status. Therefore, we will only review those studies published before 1981 and thus likely to be free from concomitant HIV infection in the study population. In 1974, 38 heroin addicts were studied in New York City.[83] Lymphocyte transformation in response to various mitogens was significantly reduced when first studied, and tended to improve on methadone maintenance, suggesting that the addicts had impairment of cell-mediated immunity. Humoral immunity was also abnormal: most had diffuse hyperglobulinemia, 23% had biologic false positive serum test for syphilis, and 21% had positive latex fixation tests. In 1977, another New York-based study showed changes in B and T cell numbers among 30 healthy addicts being maintained on methadone.[84] Although their assay was crude by current standards, they reported that B-cell numbers were increased in 40% of their population, and decreased in another 27%; T-cells were decreased in 20%. During 1978-79, 44 heroin addicts from Atlanta, Boston, and Chicago had lymphocyte studies performed, and likely represent the last addict population to be examined which was not significantly "contaminated" by coincident HIV infection.[85] The investigators found a significant reduction in patients' T-cells compared to normals, a concomitant increase in null cells, and poor lymphocyte transformation. No change in B-cells or total white blood cell count was seen. Incubation of addicts' cells with the opiate antagonist naloxone normalized both their proportion of T and null lymphocytes and their blast transformation in response to the mitogen PHA. They deduced that opiates bind reversibly to lymphocytes, blocking their T-cell antigens and their ability to transform when stimulated. As further evidence, in vitro incubation of lymphocytes with opiates has been reported to depress their blast transformation.[86] These studies suggest that opiate use may impair cell-mediated immune function at least temporarily, thereby embarrassing antiviral defense. Moreover, from the above studies plus several others,[87-89] evidence is strong that addicts have hyperactive B-cell production: hyperglobulinemia,[83,87] rheumatoid factors,[83,88] increased opsonic capacity,[88] and presence of antimuscle antibodies[89] have all been reported. The effect of this presumed chronic stimulation on normal resistance to infection is unstudied.

B. Marihuana

A large number of studies have demonstrated that tetrahydrocannabinol (THC) and other cannabinoids present in marihuana are suppressive of various cell-mediated immune functions in rodents, but data in man are less clearcut.[90,91] In vitro exposure of normal rodent and human lymphocytes to clinically achievable concentrations of cannabinoids has been shown to suppress lymphocyte transformation,[92,93] interferon production,[94] and natural killer cell activity.[95,96] In vivo administration of cannabinoids to rodents has caused depression of delayed hypersensitivity reactions,[97] lymphocyte transformation,[92] natural killer cell activity,[98] and interferon production.[99] In vivo studies in man have given inconsistent results. For example, several early reports showed reduced circulating T-lymphocytes[100] and impaired lymphocyte transformation[101] in chronic marihuana users, but others have found normal function.[102,103] A recent review summarizes human data thus: " . . . the inconsistency of the results with humans exposed to cannabinoids does not permit an interpretation that cell-mediated immunocompetence is affected."[104]

A similar situation pertains regarding humoral immunity. Although both primary and anamnestic antibody responses are inhibited by cannabinoids in rodents,[105,106] no conclusive data exists showing a similar suppression in man. Nonetheless, it is prudent to be concerned that humans might be similarly affected. To summarize, in vitro and in vivo exposure to cannabinoids significantly inhibits cell-mediated and humoral immunity in rodents. Evidence in man is less compelling, but suggests the potential for similar immunosuppression.

C. Central Nervous System Stimulants

Little data are available, none of it in man.[91] Cocaine, given to mice for 14 days, has been shown to suppress antibody response in two studies;[91,107] in contrast, one of the studies found increased delayed hypersensitivity responses[107] while the other showed no change.[91] Amphetamines given for a similar period were found to decrease the animals' total lymphocyte counts and spleen cell number, without affecting antibody responses.[91]

D. Barbiturates

In vitro incubation of murine and primate blood lymphocytes with anesthesia-inducing concentrations of various barbiturates have shown both suppression[108,109] and no effect[110,111] on their response to mitogens. But, when cells are studied from subjects given barbiturate anesthesia, significant inhibition of lymphocyte transformation, antibody-dependent cell-mediated cytotoxicity (ADCC), and natural killer cell activity are found short-term, with return to normal by 24 hours after anesthesia.[108,110] The effects of chronic barbiturate exposure in vivo, particularly in the context of addiction, have not been examined.

E. Amyl Nitrite

In 1982, Goedert and colleagues reported that homosexual men without AIDS who used amyl nitrite regularly had reversed T4:T8 lymphocyte ratios and lower absolute T4 cell counts,[112] raising the possibility that the nitrite was immunosuppressive. Subsequently however, exposure of mice to nitrites by inhalation for up to three months was reported subsequently to have no effect on lymphocyte transformation, delayed hypersensitivity, T-cell numbers, or antibody responses.[113] It now seems in retrospect that nitrite use was probably only serving as a marker for promiscuity, thereby identifying the subpopulation most likely to be infected with CMV, EBV, or other viruses known to affect T-cell proportions. At the time of Goedert's original findings, testing for HIV infection was not possible, and so his results also may have been secondary to asymptomatic HIV infection.

F. Summary

Opiate use in humans is associated with a reduction in the number of circulating T-lymphocytes and in their ability to undergo transformation, probably resulting from opiate binding to T-cell antigens. (See Table 4.) Patients also manifest a diffuse hyperglobulinemia, without change in circulating B-cell numbers. In vitro exposure of rodent and human lymphocytes to cannabinoids depresses

Table 4

Effects of Various Drugs on Immune Functions

Drug	Effects
Opiates	Decreased # circulating T-lymphocytes; ↑ null cells. Decreased lymphocyte transformation. Diffuse hyperglobulinemia
Marihuana	In rodents: decreased LT, NK activity, interferon production, delayed hypersensitivity and antibody responses. Data in man are mixed.
CNS Stimulants	Data inadequate
Barbiturates	Acute effects: ↓ LT, NK activity, ADCC.
Amly Nitrite	No consistent findings

their transformation, natural killer activity, and interferon production. Similar results occur with in vivo exposure of rodents, but data in man are mixed. Cannabinoids also inhibit primary and secondary antibody responses in rodents. Data regarding the effects of CNS stimulants are too scanty to allow comment. Barbiturate anesthesia in man and animals produces a short-lived depression of lymphocyte transformation, NK activity, and ADCC; no studies of the effects of chronic exposure have been performed. Amyl nitrite has not been proven to possess any immunosuppressive activity.

NOTES

1. Sissons JGP, Oldstone MBA. Host response to viral infections. In: Fields BN et al., ed. Virology. New York: Raven Press, 1985:265-79.

2. Dalgleish AG, Beverley PCL, Clapham PR, Crawford DH, Greaves MF, Weiss RA. The CD4 (T4) antigen is an essential component of the receptor for the AIDS retrovirus. Nature. 1985; 312:763-7.

3. Schwartz K, Visscher BRT, Detels R, Taylor J, Nishanian P, Fahey JL. Immunological changes in lymphadenopathy virus positive and negative symptomless male homosexuals: two years of observation. Lancet 1985; 2:831-2.

4. MacGregor RR. Alcohol and immune defense. JAMA. 1986; 256:1474-9.

5. Adams HG, Jordan C. Infections in the alcoholic. Med Clin North Am. 1984; 68:179-200.

6. Smith FE, Palmer DL. Alcoholism, infection, and altered host defenses. J Chronic Dis. 1976; 29:35-49.

7. Koch R. "Uber die cholerabahterien." G Riemer, Berlin, 1884.

8. Capps JA, Coleman GH. Influence of alcohol on prognosis of pneumonia in Cook County Hospital. JAMA. 1923; 80:750-2.

9. Nolan JP. Alcohol as a factor in the illness of university service patients. Am J Med Sci. 1965; 249:135-42.

10. Schmidt W, DeLint J. Causes of death of alcoholics. Q J Stud Alcohol. 1972; 33:171-85.

11. Bartlett JG, Finegold SM. Anaerobic infections of the lung and pleural space. Am Rev Respir Dis. 1974; 110:56-77.

12. Conn HO, Fessel JM. Spontaneous bacterial peritonitis in cirrhosis: Variations on a theme. Medicine. 1971; 50:161-97.

13. Snyder N, Bassoff J, Dwyer, et al. Depressed cutaneous delayed hypersensitivity in alcoholic hepatitis. Digest Dis. 1978; 23:353-8.

14. Berenyi MR, Straus B, Cruz D. In vitro and in vivo studies of cellular immunity in alcoholic cirrhosis. Am J Dig Dis. 1974; 119:199-205.

15. Bjorkholm M. Immunological and hematological abnormalities in chronic alcoholism. Acta Med Scand. 1980; 207:197-200.

16. Spagnuolo PJ, MacGregor RR. Acute ethanol effect on chemotaxis and other components of host defense. J Lab Clin Med, 1975; 86:24-31.

17. Gluckman SJ, Dvorak VC, MacGregor RR. Host defenses during prolonged alcohol consumption in a controlled environment. Arch Intern Med 1977; 137:1539-43.

18. Roselle GA, Mendenhall CL. Ethanol-induced alterations in lymphocyte function in the guinea pig. Alcoholism. 1984; 8:62-7.

19. Lundy J, Raaf, JH, Deakins S, et al. The acute and chronic effects of alcohol on the human immune system. Surg Gynecol Obstet. 1975; 141:212-8.

20. Long JM, Ruscher H, Hasselmann JP, et al. Decreased autologous rosette-forming T-lymphocytes in alcoholic cirrhosis. Int Arch Allergy Appl Immunol. 1980; 61:337-43.

21. Tennenbaum JI, Ruppert RD, St Pierre RL, et al. The effect of chronic alcohol administration on the immune responsiveness of rats. J Allergy Clin Immunol. 1969; 44:272-81.

22. Eichner ER, Buchanan B, Smith JW, et al. Variations in the hematologic and medical status of alcoholics. Am J Med Sci. 1972; 263:35-42.

23. Liu YK. Effects of alcohol on granulocytes and lymphocytes. Semin Hematol. 1980; 17:130-6.

24. McFarland W, Libre EP. Abnormal leukocyte response in alcoholism. Ann Intern Med. 1963; 59:865-77.

25. Bernstein IM, Webster KH, Williams RC, et al. Reduction in circulating T-lymphocytes in alcoholic liver disease. Lancet. 1974; 2:488-90.

26. Berenyi MR, Straus B, Avila L. T rosettes in alcoholic cirrhosis of the liver. JAMA. 1975; 232:44-6.

27. Tisman G, Herbert V. In vitro myelosuppression and immunosuppression by ethanol. J Clin Invest. 1973; 51:1410-4.

28. Roselle Ga, Mendenhall CL. Alteration of in vitro human lymphocyte function by ethanol, acetaldehyde, and acetate. J Clin Lab Immunol. 1982; 9:33-7.

29. Glassman AB, Bennett CE, Randall CL. Effects of ethyl alcohol on human peripheral lymphocytes. Arch Pathol Lab Med. 1985; 109:540-2.

30. Sorrell MF, Leevy CM. Lymphocyte transformation and alcoholic liver injury. Gastroent. 1972; 63:1020-5.

31. Kawanishi H, Tavassolie H, MacDermott P, et al. Impaired concanavalin A-inducible suppressor T-cell activity in alcoholic liver disease. Gastroent. 1981; 80:510-7.

32. Hsu CCS, Leevy CM. Inhibition of PHA-stimulated lymphocyte transformation by plasma from patients with advanced alcoholic cirrhosis. Clin Exp Immunol. 1971; 8:749-60.

33. Young GP, Van der Weyden MB, Rose IS, et al. Lymphopenia and lymphocyte transformation in alcoholics. Experimentia. 1978; 35:267-9.

34. Rice C, Hudig D, Lad P, et al. Ethanol activation of human natural cytotoxicity. Innumopharm. 1983; 6:303-16.

35. Ristow SS, Starkey JR, Hass GM. Inhibition of natural killer cell activity in vitro by alcohols. Biochem Biophys Res Commun. 1982; 105:1315-21.

36. Saxena QB, Mezey E, Adler WH. Regulation of natural killer activity in vivo: The effect of alcohol consumption on human peripheral blood natural killer activity. Int J Cancer. 1980; 26:413-7.

37. Abdallah RM, Starkey JR, Meadows GG. Alcohol and related dietary effects on mouse natural killer-cell activity. Immunology. 1983; 50:131-7.

38. Charpentier B, Franco D, Paci K, et al. Deficient natural killer cell activity in alcoholic cirrhosis. Clin Exp Immunol. 1984; 58:107-15.

39. Stacey NH. Inhibition of antibody-dependent cell-mediated cytotoxicity by ethanol. Immunopharm. 1984; 8:155-61.

40. Kaelin RM, Semerjian A, Center DM, et al. Influence of ethanol on human T-lymphocyte migration. J Lab Clin Med. 1984; 104:752-60.

41. Atkinson JP, Sullivan TJ, Kellyn JP, et al. Stimulation by alcohols of cyclic AMP metabolism in human leukocytes. J Clin Invest. 1977; 60:284-94.

42. Green GM, Kass EH. Factors influencing the clearance of bacteria by the lung. J Clin Invest. 1964; 43:769-76.

43. Okeson GC, Divertie MB. Cilia and bronchial clearance: The effects of pharmacologic agents and disease. Mayo Clin Proc. 1970; 45:361-73.

44. Bomalaski JS, Phair JP. Alcohol, immunosuppression, and the lung. Arch Intern Med. 1982; 142:2073-4.

45. Astry CL, Warr GA, Jakab GJ. Impairment of PMN immigration as a mechanism of alcohol-induced suppression of pulmonary antibacterial defenses. Am Rev Respir Dis. 1983; 128:113-7.

46. Guarneri JJ, Laurenzi GA. Effect of alcohol on the mobilization of alveolar macrophages. J Lab Clin Med. 1968; 72:40-51.

47. Rimland D, Hand WL. The effect of ethanol on adherence and phagocytosis by rabbit alveolar macrophages. J Lab Clin Med. 1980; 95:918-26.

48. Rimland D. Mechanisms of ethanol-induced defects of alveolar macrophage function. Alcoholism. 1983; 8:73-6.

49. Louria DB. Susceptibility to infection during experimental alcoholic intoxication. Trans Assoc Am Physicians. 1963; 76:102-10.

50. Galente D, Andreana A, Perna P, et al. Decreased phagocytic and bactericidal activity of the hepatic reticuloendothelial system during chronic ethanol treatment, and its restoration by levamisole. J Reticuloendothel Soc. 1982; 32:179-87.

51. Ali MV, Nolan JP. Alcohol induced depression of reticuloendothelial function in the rat. J Lab Clin Med. 1967; 70:295-301.

52. Liu YK. Phagocytic capacity of RES in alcoholics. J Reticuloendothel Soc. 1979; 25:605-13.

53. Gilhus NE, Matre R. In vitro effect of ethanol on subpopulations of human blood mononuclear cells. Int Arch Allergy Appl Immunol. 1982; 68:382-6.

54. Parkinson PR. The relation of alcohol to immunity. Lancet. 1909; 2:1580.

55. Loose LD, Stege T, Diluzio NR. The influence of acute and chronic

ethanol or bourbon administration on phagocytic and immune responses in rats. Exp Mol Pathol. 1975; 23:159-72.

56. Bjorneboem M, Jensen KB, Scheibel I, et al. Tetanus antitoxin production and gamma globulin levels in patients with cirrhosis of the liver. Acta Med Scand. 1970; 188:541-6.

57. Pirovino M, Lydick E, Grob PJ, et al. Pneumococcal vaccination: The response of patients with alcoholic liver cirrhosis. Hepatology. 1984; 4:946-9.

58. Smith WI, Van Thiel DH, Whiteside T, et al. Altered immunity in male patients with alcoholic liver disease. Alcoholism. 1980; 4:199-206.

59. MacGregor RR, Gluckman SJ, Senior JR. Granulocyte function and levels of immunoglobulins and complement in patients admitted for withdrawal from alcohol. J Infect Dis. 1978; 138:747-53.

60. Wilson ID, Onstad G, Williams RC. Serum immunoglobulin concentration in patients with alcoholic liver disease. Gastroent. 1969; 57:59-67.

61. Wands JR, Dienstag JL. Weake JR, et al. Proliferative and secretory B cell activity in severe alcoholic liver disease. Gastroent. 1978; 75:992.

62. Rong PB, Kalsi J, Hodgson JF. Hyperglobulinaemia in chronic liver disease: Relationships between in vitro immunoglobulin synthesis, short-lived suppressor cell activity and serum immunoglobulin levels. Clin Exp Immunol. 1984; 55:546-52.

63. Thomas HC, McSween RNM, White RG. Role of the liver in controlling the immunogenicity of commensal bacteria in the gut. Lancet. 1973; 1:1288-91.

64. Bjorneboem M, Brytz H. Antibodies to intestinal microbes in serum of patients with cirrhosis of the liver. Lancet. 1972; 1:58-60.

65. Grieco MN, Capra JD, Paderon H. Reduced serum beta 1c/1a globulin levels in extrarenal disease. Am J Med. 1971; 51:340-5.

66. Johnson WD, Stokes P, Kaye D. The effect of intravenous ethanol on the bactericidal activity of human serum. Yale J Biol Med. 1969; 42:71-85.

67. Kaplan NM, Braude AI. Hemophilus influenzae infection in adults. Arch Intern Med. 1958; 101:515-23.

68. Marr JJ, Spilberg I. A mechanism for decreased resistance to infection by gram-negative organisms during acute alcoholic intoxication. J Lab Clin Med. 1975; 86:253-8.

69. Imperia PS, Chikkappa G, Phillips PG. Mechanism of inhibition of granulopoiesis by ethanol. Proc Soc Exp Biol Med. 1984; 175:219-25.

70. Pickrell KL. The effect of alcoholic intoxication and ether anesthesia on resistance to pneumococcal infection. Bull Johns Hopkins Hosp. 1928; 63:238-60.

71. Brayton RG, Stokes PE, Schwartz MS, Louria D. Effect of alcohol and various disease on leukocyte mobilization, phagocytosis and intracellular bacterial killing. N Engl J Med. 1970; 282:123-8.

72. MacGregor RR, Gluckman SG. Effect of acute alcohol intoxication on granulocyte mobilization and kinetics. Blood. 1979; 52:551-9.

73. MacGregor RR, Spagnuolo PJ, Lentnek AL. Inhibition of granulocyte

adherence by ethanol, prednisone, and aspirin measured with a new assay system. N Engl J Med. 1974; 291:642-6.

74. Buckley RM, Ventura ES, MacGregor RR. Propranolol antagonizes the anti-inflammatory effect of alcohol and improves survival of infected intoxicated rabbits. J Clin Invest. 1978; 62:554-9.

75. DeMeo AN, Andersen BR. Defective chemotaxis associated with a serum inhibitor in cirrhotic patients. N Engl J Med. 1972; 286:735-40.

76. Van Epps DE, Strickland RG, Williams RC. Inhibitors of leukocyte chemotaxis in alcoholic liver disease. Ann Intern Med. 1975; 59:200-7.

77. Maderazo EG, Ward PA, Quintiliani R. Defective regulation of chemotaxis in cirrhosis. J Lab Clin Med. 1975; 85:621-30.

78. Feliu E, Gougerot MA, Hakim J, et al. Blood polymorphonuclear dysfunction in patients with alcoholic cirrhosis. Eur J Clin Invest. 1977; 7:571-7.

79. MacGregor RR. In vitro chemotaxis and in vivo delivery of PMNs in men with alcoholic cirrhosis. Clin Res. 1986; 34:524.

80. Hallengren B, Forsgren A. Effect of alcohol on chemotaxis, adherence, and phagocytosis of human PMNs. Acta Med Scand. 1978; 204:43-8.

81. Orangio GR, Latta PD, Marino C, Guarneri JJ, Giron JA, Palmer J, Margolis IB. Infections in parenteral drug abusers. Amer J Surg. 1983; 146:738-41.

82. Lefkowitz SS, Chiang CY. Effects of certain abused drugs on hemolysin forming cells. Life Sci. 1975; 17:1763-8.

83. Brown SM, Stimmel B, Taub RN, Kochwa S, Rosenfield RE. Immunologic dysfunction in heroin addicts. Arch Intern Med. 1974; 134:1001-6.

84. Cushman P, Gupta S, Grieco MH. Immunological studies in methadone maintained patients. Internat J Addict. 1977; 12:241-53.

85. McDonough RJ, Madden JJ, Falek A, Shafer DA, Pline M. Gordon D, Bokos P. Kuehnle JC, Mendelson J. Alteration of T and null lymphocyte frequencies in the peripheral blood of human opiate addicts: in vivo evidence for opiate receptor sites on T lymphocytes. J Immunol. 1980; 125:2539-43.

86. Saito K, Kumagai K. The suppressive effect of meperidine on PHA-stimulated transformation of human lymphocytes. Tohaku J Exp Med. 1981; 134:337-8.

87. Cushman P, Grieco MH. Hyperimmunoglobulinemia associated with narcotic addiction. Amer J Med. 1973; 54:320-6.

88. Nickerson DS, Williams RC, Boxmeyer M, Quie PG. Increased opsonic capacity of serum in chronic heroin addiction. Ann Intern Med. 1970; 72:671-7.

89. Husby G, Pierce PE, Williams RC. Smooth muscle antibody in heroin addicts. Ann Intern Med. 1975; 83:801-5.

90. Munson AE, Fehr KO. Immunologic effects of cannabis. In: Fehr KO and Kalant H, ed. Adverse Health and Behavioral Consequences of Cannabis Use, New York: Raven Press, 1983: 257-63.

91. Holsapple MP, Munson AE. Immunotoxicology of Abused Drugs. In: J Dean et al., ed. Immunotoxicology and Immunopharmacology, New York: Raven Press, 1985: 381-92.

92. Pruess MM, Lefkowitz SS. Influence of maturity on immunosuppression by tetrahydrocannabinol. Proc Soc Exp Biol Med. 1978; 158:350-3.

93. Klein TW, Newton CA, Widen R, Friedman H. The effect of THC and 11-hydroxy-THC on T-lymphocyte and B-lymphocyte mitogen responses. J Immunopharm. 1985; 7:451-66.

94. Blanchard DK, Newton C, Klein TW, Stewart WE, Friedman H. In vitro and in vivo suppressive effects of THC on interferon production by murine spleen cells. Int J Immunopharmac. 1986; 8:819-24.

95. Specter SC, Klein TW, Newton C, Mondragon M, Widen R, Friedman H. Marijuana effects on immunity: Suppression of human natural killer cell activity by THC. Int J Immunopharmac. 1986; 8:741-5.

96. Klein TW, Newton C, Friedman H. Inhibition of natural killer cell function by marijuana components. J Toxicol Environ Health. 1987; 20: (in press).

97. Morahan PS, Klykken PC, Smith SH, Harris LS, Munson AE. Effects of cannabinoids on host resistance to Listeria monocytogenes and herpes simplex virus. Infect Immun. 1979; 23:670-4.

98. Patel V, Borysenko M, Kumar MSA, Millard WJ. Effects of acute and subchronic THC administration on the plasma catecholamine, beta-endorphin, and corticosterone levels and splenic natural killer activity in rats. Proc Soc Exp Biol Med. 1985; 180:400-4.

99. Cabral GA, Lockmuller JC, Mishkin EM. THC decreases alpha/beta interferon response to herpes simplex virus type 2 in the B6C3F1 mouse. Proc Soc Exp Biol Med. 1986; 181:305-11.

100. Gupta S, Grieco MH, Cushman P. Impairment of rosette-forming T-lymphocytes in chronic marihuana smokers. N Engl J Med. 1974; 291:874-7.

101. Lau RJ, Tubergen DG, Barr M, Domino EF, Benowitz N, Jones RT. Phytohemagglutinin-induced lymphocyte transformation in humans receiving THC. Science. 1976; 192:805-7.

102. Kaklamani E, Trichopoulos D, Koutselinis A, Drouga M. Hashish smoking and T-lymphocytes. Arch Toxicol. 1978; 40:97-101.

103. White SC, Brin SS, Janicki BW. Mitogen-induced blastogenic responses of lymphocytes from marihuana smokers. Science. 1975; 188:71-2.

104. Munson AE, Holsapple MP: Overview of the immunotoxicology of marihuana. In: Harvey DJ, ed. Marihuana '84. Proceedings of the Oxford Symposium on Cannabis, Oxford: IRL Press Limited, 1984:419-30.

105. Smith SH, Harris LS, Uwaydah IM, Munson AE. Structure-activity relationships of natural and synthetic cannabinoids in suppression of humoral and cell-mediated immunity. J Pharmacol Exp Ther. 1978; 207:165-70.

106. Zimmerman S, Zimmerman AM, Cameron IL, Laurence HL. THC, cannabidiol, and cannabinol effects on the immune response of mice. Pharmacol. 1977; 15:20-3.

107. Faith RE, Valentine JL. Effects of cocaine exposure on immune function. Toxicol. 1983; 3:56.

108. Formeister JF, MacDermott RP, Wickline D, Locke D, Nash GS, Rey-

nolds DG, Robertson BS. Alteration of lymphocyte function due to anesthesia. Surgery 1980; 87:573-80.

109. Neuwelt EA, Kikuchi K, Hill SA, Lipsky P, Frenkel E. Barbiturate inhibition of lymphocyte function. J Neurosurg. 1982; 56:254-9.

110. Thomas J, Carver M. Haisch C, Thomas F, Welch J, Carchman R. Differential effects of intravenous anesthetic agents on cell mediated immunity in the rhesus monkey. Clin Exp Immunol. 1982; 47:457-66.

111. Gabourel JP, Davies GH, Rittenberg MB. Effects of salicylate and phenobarbital on lymphocyte proliferation and function. Clin Immun Immunopath. 1977; 7:53-61.

112. Goedert JJ, Wallen WC, Mann DL, Strong DM, Neuland CY, Greene MH, Murray C, Fraumeni JF, Blattner WA. Amyl nitrite may alter T lymphocytes in homosexual men. Lancet 1982; 1:412-6.

113. Lewis DM, Koller WA, Lynch DW, Spira TJ. Absence of immunotoxic effects of inhaled isobutyl nitrite in Balb/c mice. Toxicologist. 1984; abstract 427.

The Prevention
of HIV Infection Associated
with Drug and Alcohol Use
During Sexual Activity

Ron Stall, PhD, MPH

SUMMARY. The effort to prevent further HIV infection for high risk populations has been largely limited, in practice, to health education efforts. Prevention policy must utilize techniques which, in combination with health education efforts, works to decrease behaviors which are implicated in HIV transmission. It is argued that this approach to prevention will work best if it is designed with an understanding of the conditions under which individuals decline to comply with risk-reduction guidelines. It appears that the use of drugs and alcohol during sexual contact is one such condition. Three prevention strategies are suggested which, in addition to health education efforts, might be used to minimize HIV transmission related to the use of drugs and alcohol during sexual activity among gay men. It is hoped that the approach suggested in this paper for the preven-

Ron Stall, PhD, MPH, The Center for AIDS Prevention Studies, University of California, San Francisco, CA 94143.

This paper could not have been written without the participation of the sampled volunteers for the San Francisco Men's Health Study as well as the AIDS Behavioral Research Study. The men followed for these prospective studies have patiently responded to questions regarding intimate lifestyle issues for a number of years in the hope that AIDS epidemic might be brought to an end. Data described in this paper were generated by studies supported by the National Institute of Allergy and Infectious Diseases (The San Francisco Men's Health Study, A1-32519) and by the National Institute of Mental Health (The AIDS Behavioral Research Study, MH 39553). Support for this paper was provided in part through a center grant from the National Institute of Mental Health and the National Institute of Drug Abuse (The Center for AIDS Prevention Studies, P50 MH42459).

73

tion of further HIV infection might also be adopted to develop prevention strategies for other populations at risk of HIV infection.

INTRODUCTION

As has been the case since the earliest days of the AIDS epidemic, prevention of further HIV infection has been our only tool to control this epidemic. Typically, AIDS prevention strategies have relied most heavily on health education efforts. This widespread policy has been based on the assumption that informing individuals of the risk associated with certain behaviors will cause a discontinuance of these behaviors. However, survey research designed to measure changes in risk behaviors among gay men has demonstrated that a significant minority within this population has continued to engage in high risk behaviors despite a widespread understanding of the specific behaviors which facilitate HIV infection.[1] It can also be assumed that noncompliance with well-known risk reduction guidelines exists among addicted needle users. Perhaps the most important lesson which we have learned so far from the AIDS prevention literature is that health education is a necessary but not sufficient cause for universal risk-related behavioral change. This lesson should surprise no one: it was also learned in efforts to control smoking and to encourage drivers to wear seatbelts.

The current challenge to those interested in the prevention of further HIV infection thus lies in the identification and implementation of prevention strategies which, in combination with health education efforts, will serve as sufficient causes for further AIDS risk reduction. For example, AIDS prevention efforts might go beyond the dissemination of risk reduction information to nonobtrusively modify the environment in which risk-taking occurs. That is, condoms could be made easily available in "pick-up" bars or small bottles of bleach could be given to intravenous users so that if needles are shared they can at least be sterilized. It is asserted here that the probability of success for such prevention/health education strategies will be increased if they are designed to be responsive to the conditions under which it has been shown that individuals decline to comply with risk reduction guidelines. Thus, understanding the correlates of high risk behavior is a necessary first step towards the creation of an effective AIDS prevention policy.

CORRELATES OF
HIGH RISK SEXUAL ACTIVITY:
FINDINGS FROM A GROWING LITERATURE

A small, but growing, survey research literature has identified a set of correlates of high risk behavior among gay men. As this literature expands to include heterosexual and high risk minority populations as well as building on the small survey literature on intravenous drug users, our chances to design effective prevention policy will increase. Reflecting the emphasis of the literature, the remainder of this paper will be restricted to the discussion of strategies for the prevention of sexual transmission of HIV among gay men. However, it is hoped that the ideas proposed here might be used to stimulate prevention policy and research specific to the circumstances of other populations at high risk.

Although surveys of the behavioral changes made by gay men have identified a set of characteristics associated with high risk sexual activity, no attempt to unite these variables into an explanatory model of risk taking behavior has yet been published. For this reason, variables associated with high risk sexual behavior can only be listed—little understanding of the explanatory power or theoretical importance of any particular variables has yet been achieved.

For example, Weber, Coates and McKusick[2] found that denial of personal risk for the onset of AIDS was associated with high risk sexual behavior. This is consistent with a similar finding from the same data set: men who believed that they had been exposed to, but had fought off, an HIV infection reported significantly greater numbers of sexual partners and more risky sex than those who believed that they were incubating the virus or those who believed that they had not been exposed.[3] Stall, et al.[4] found a strong relationship between the use of drugs and/or alcohol during sexual activity and noncompliance with safe sex techniques. Significantly more potentially unsafe sexual activity occurs within primary gay relationships than between men who are not as seriously involved,[1] although if the relationship is monogamous no further HIV infection can result.

The cohort of gay men currently older than 40 have higher rates of participation in high risk sexual activity than do younger men, however it also appears that older gay men are more likely to cut back on high risk sexual activity over time than are younger age groups.[5] Conversely, although gay men in their twenties were not especially likely to be at high risk in a cross-sectional analysis, they were significantly less likely to sustain low levels of sexual risk than were older gay men. Men who hold a visual image of the physical deterioration which accompanies AIDS have reduced numbers of sexual partners.

A recent analysis found that five variables were significantly related to sexual risk in a multiple regression analysis of longitudinal data.[1] These included, in descending order of importance, personal efficacy (that is, the belief that one is capable of making recommended changes); relationship status (those in relationships engage in high risk sex more frequently); depression (associated with lowered risk); relative youth (gay men in their twenties were less likely to sustain low levels of risk over time); and level of agreement with AIDS risk reduction guidelines (those most likely to agree sustain low levels of risk).

Although these analyses are quite useful for designing specialized health education campaigns, those who hope to use this literature to guide prevention beyond health education strategies must find this a somewhat frustrating list of variables. That is, the majority of these variables are relatively fixed and therefore not amenable, for the most part, to manipulation with the goal of minimizing HIV transmission. It seems reasonable to expect that increasing the prevalence of the psychological characteristic of personal efficacy within a population could only be accomplished at considerable cost and effort, as would decreasing the rates of personal denial of the threat of AIDS. The characteristic of age is not manipulable by public health practitioners, nor is relationship status. Using media techniques to portray the physical deterioration of AIDS victims may induce panic, and it can hardly be recommended that efforts be made to increase depression within a population so that risk for HIV transmission could be lowered. Further, depression could in fact be reactive to knowing someone with AIDS, itself associated with lowered risk. For these reasons, understanding the associations between high risk sexual activity and drug and alcohol use during sexual activity may be enormously important since this particular combination of behaviors is uniquely appropriate to manipulation through prevention strategies.

THE ASSOCIATIONS BETWEEN HIGH RISK SEXUAL ACTIVITY AND SUBSTANCE USE DURING SEXUAL ACTIVITY

Data described here were drawn from the May, 1984 and May, 1985 waves of data collection for the AIDS Behavioral Research Project (n = 463). As this analysis (including a detailed description of the methodology) have been reported elsewhere,[3,4] this section will report findings as they are applicable to the prevention of further HIV infection.

Table 1 describes the association between using drugs or alcohol during sexual activity and a summary score which measures sexual risk for HIV transmission. Percentage figures are only given for the men who combined a particular drug with sexual activity at each risk level. Thus, subtracting the figures in each cell from 100% gives the proportion of men at each risk level who did not use a particular drug during sexual activity over the past 30 days. Depending on the specific drug, the men at high risk are from 2 to 3.5

TABLE 1: Use of Drugs During Sexual Activity by AIDS Risk Scores, In Percentages

	Drug Used During Sex Last Month			
	Alcohol (n=298)	Poppers (n=202)	Marijuana (n=218)	"Other Drugs" (n-101)
Risk Score for AIDS				
No Risk	38.5	25.6	25.6	10.3
Medium Risk	66.7	39.2	42.4	15.8
High Risk	79.4	66.0	65.7	35.4
Chi Square=	52.0	35.8	47.5	31.8
p significance	.0000	.0000	.0000	.0000

times more likely to have used drugs during sexual activity than the men at no risk. The men at medium risk are approximately 1.5 times more likely to have used drugs during sexual activity than the men at no risk. In each case, the differences in probability of drug use by risk category are highly significant. The proportionate increase in risk for risky sexual practices appears to be greater if the drug used during such activity is illegal (e.g., marijuana and "other drugs") rather than legal and easily available (e.g., alcohol and poppers). It is also important to point out that except in the case of "other drugs," the majority of men at high risk combined drug use with sexual activity at least once during the previous month.

It is also important to point out that similar relationships were noted in the relationship between total number of drugs used during sex and high risk sex. Further, the higher the proportion of high risk sexual encounters for all sexual encounters, the more likely drug use was combined with sexual activity. Together these findings indicate that there is a strong cross-sectional relationship between high risk sexual activity and use of drugs during sexual encounters. It appears that use of drugs and/or alcohol during sexual contacts is one condition under which individuals sometimes decline to comply with risk reduction guidelines.

Of considerable interest to prevention concerns is the question of whether changing risky sexual behavior is related to combining drugs and/or alcohol use with sexual activity. Table 2 summarizes the relationship between drug use during sexual activity and change in risk scores from May, 1984 to May, 1985. This table compares men who were at no risk as of May, 1984 to men at high risk at the same time (n = 311). By May of 1985, some of the men who were originally at no risk increased their risk scores for HIV infection, similarly, some men who originally had high risk scores decreased risky sexual behavior. Each of these four distinct groups is compared according to current drug use behavior during sexual activity (e.g., as of May, 1985). These comparisons allow a measure of whether men who currently use drugs or alcohol during sexual activity had different histories of having adopted safe sex guidelines over the previous year.

In Table 2 the proportion of men who used drugs at least once during sexual activity during the first wave of data collection is

TABLE 2: Prevalence of Use of Specific Drugs During Sex in May, 1985 by Change in Risk Scores from May 1984-May 1985

	Used During Sex Last Month:			
	Alcohol	Poppers	Marijuana	"Other Drugs"
No Risk -- May, 1984				
T1 No Risk T2 No Risk (n=46)	32.6	17.4	20.0	4.3
T1 No Risk T2 More Risk (n=54)	77.8	37.0	46.3	29.6
High Risk -- May, 1984				
T1 High Risk T2 Less Risk (n=95)	55.8	49.0	40.6	15.6
T1 High Risk T2 High Risk (n=116)	75.9	64.7	67.2	36.2

compared according to change in risk scores during the ensuing year. The men who were originally at no risk but who increased risk upon follow-up (row 2) were at least 2 times more likely to use alcohol, poppers or marijuana during sexual activity than were the men who remained at no risk (row 1). Furthermore, men who increased risk were nearly 7 times more likely to use "other drugs" than the men who remained at no risk. Similar findings emerge in a comparison of the men originally at high risk but who decreased risky sexual behavior (row 3) and those who remained at high risk (row 4). The men who remained at high risk were approximately 1.5 times more likely to use alcohol, poppers or marijuana during sexual activity than were the men who decreased risk.

Also of interest in Table 2 is a comparison of the men who remained at no risk with the men who remained at high risk. The men who remained at high risk were 2.3 times more likely to use alcohol during sexual activity, 3.7 times more likely to use poppers, 3.4 times more likely to smoke marijuana, and 8.4 times more likely to use "other drugs." Together these findings form a consistent pattern: men originally at no risk, but who increased risk over the previous year, are more likely to use drugs and alcohol during sex-

ual activity than were the men who remained at no risk. Men originally at high risk for HIV infection, yet who managed to decrease
risk over the past year, are less likely to use drugs and alcohol
during sexual activity than the men who remained at high risk. Further, change or stability in high risk sexual practices over the past
year is associated in a similar way with the number of different
drugs used during sex and with the proportion of sexual encounters
involving drugs. In both cases, low drug use is related to the maintenance of low risk sex practices and to switching from high to low
risk behaviors.

Although it may seem that the relationship between drug and/or
alcohol use during sexual activity is rather simple and to be expected, numerous hypotheses can be forwarded to explain this association and each of these hypotheses have very different prevention
implications. These hypotheses can be labelled the disinhibition,
aphrodisiac, personality, social context, multifactorial and null hypotheses. The theoretical and health policy implications of these
competing hypotheses are described in detail elsewhere.[4,6] However, it is clear that whatever the cause behind this association, the
strength of the association between combining sexual activity with
drug and/or alcohol use and high risk sexual behavior is so strong
and the consequences of infection with HIV so profound that health
education campaigns to communicate the fact of this association to
populations at high risk of HIV infection seem justified. Further
prevention efforts, which might be attempted in combination with
health education campaigns, are outlined later in this paper.

THE EPIDEMIOLOGY OF DRUG AND ALCOHOL
USE AMONG GAY MALES

Given that a strong association between drug and/or alcohol use
during sexual activity and high risk sexual behavior exists, the
question can be raised whether it makes more sense in terms of
AIDS prevention efforts to attempt to prevent substance use among
gay men or whether it would be more effective to attempt to prevent
the *combination* of drug and/or alcohol use and sex. This question is
justified since a small body of literature drawn from nonrandom
samples of gay men indicates that very high prevalence rates for

alcoholism exist within this community.[7-9] Thus it could be argued that if strategies to prevent alcohol consumption were attempted within the gay male community, a set of desirable health-related outcomes would result, relating both to decreases in alcohol consumption and HIV infection rates.

However, the literature which has reported high prevalence rates for alcoholism within the gay male community has some obvious methodological shortcomings,[10] primarily having to do with the fact that samples of gay men have been drawn opportunistically and typically without heterosexual controls. This section will report findings which will determine whether the unusually high prevalence rates for heavy drug and alcohol use already reported in the literature for gay men is also observed in data drawn from a large scale random household sample of homosexual and heterosexual men living within an urban district of San Francisco. In addition, findings will be reported so that it can be determined whether homosexual and heterosexual men drink or use drugs differently from each other.

Data regarding drug and alcohol use patterns for homosexual and heterosexual men are drawn from the San Francisco Men's Health Study (SFMHS), a three-year prospective study of a cohort of single men 25-54 years of age who live in the 19 census tracts of San Francisco where the AIDS epidemic has been concentrated. The purpose of the SFMHS is to elucidate the natural history of AIDS; drug and alcohol use patterns were measured due to the widespread suspicion that recreational substance use plays a role in the spread and/or clinical manifestation of AIDS.

The 19 census tracts which define the sampling boundaries for the SFMHS cohort are clustered around the Castro district and contained some 81,291 residents as of 1980, about 12% of the city's population at that time. Approximately 23,000 (about 29%) of the residents of these census tracts satisfied the eligibility requirements for participation in the SFMHS: being male, between the ages of 25-54 and currently unmarried. The cohort was obtained by multistage cluster sampling with households as the primary sampling units and blocks and census tracts as the strata. A sample size of 1,034 men was recruited, representing a participation rate of 56% of the contacted eligibles. To adjust the prevalence estimates for difference in response probabilities by block of residence, special weights were used in addition to the cluster weights generated by the sample selection process. Comparisons of sample estimates with census figures on age distribution and AIDS surveillance data suggest that the weighted sample estimates are not seriously biased.[11] The data reported here are weighted.

The measures of sexual orientation and drug and alcohol use patterns all rely upon self-reported data. Sexual orientation was measured by utilizing a seven point scale, ambisexual males who indicated that their primary sexual attraction was to women (n = 29) were analysed as part of the gay male sample. The quantity and frequency scale was constructed according to a strategy devised by Cahalan, Roizen and Room.[12] Frequency of use of a wide variety of drugs was also measured.

Table 3 gives the prevalence of each level of quantity-frequency of typical alcohol consumption from abstention to frequency/heavy drinking for homosexual and heterosexual men. The distributions between these two groups of men are not significantly different. The modal category for both groups of men is at the frequent/light category. Although gay men are 2.2 times more likely to be abstainers than are heterosexual men, this is not a statistically significant difference. It may be, however, that this somewhat increased preva-

TABLE 3: Quantity/Frequency of Alcohol Consumption During the Past Six Months by Sexual Preference, In Percentages

	Sexual Preference	
	Homosexual (n=817)	Heterosexual (n=211)
Drinking Pattern		
Abstainer	5.6	2.5
Occasional	5.6	6.7
Infrequent	13.4	14.2
Frequent/Light	56.6	65.2
Frequent/Heavy	18.8	11.3

Mann-Whitney U Significance: .33, ns

lence of abstention reflects a larger population of recovering alcoholics who do not drink among the gay male population.

Despite the fact that there are remarkably few differences in this sample of heterosexual and homosexual men in drinking practices, important differences exist between these two groups of men in prevalence of drug use during the previous six months. From Table 4 it can be seen that gay men are significantly more likely to use marijuana, poppers, MDA, psychedelics, barbiturates, ethyl chloride and amphetamines. Due to the fact that popper use is nearly nonexistent among heterosexual men, gay men are approximately 58 times more likely than heterosexual men to use poppers during the past six months. Further, gay men are five times more likely to use MDA, three times more likely to use barbiturates, and two times more likely to use amphetamines (other than MDA) than are heterosexual men. Proportionate differences in the use of marijuana, cocaine, psychedelics and the opiates are not large between the two groups. The relative similarity in prevalence of use of the

TABLE 4: Prevalence of Use of Specific Drugs During the Past Six Months by Sexual Preference, In Percentages

Drug Used	Sexual Preference	
	Homosexual (n=817)	Heterosexual (n=211)
Marijuana*	77.5	70.6
Poppers+	57.7	1.0
Cocaine	47.9	49.2
MDA+	8.5	1.6
Psychedelics*	17.5	12.5
Barbiturates+	24.9	8.5
Ethyl Chloride+	2.8	--
Opiates	4.0	4.9
Amphetamines+	28.2	17.2

Chi Square significance=   *p less than .05
                                     +p less than .01

drugs listed in Table 3, in comparison to poppers, suggests that if there is a distinctively "gay drug," that drug is poppers.

However, if the prevalence of *weekly* use of drugs is compared between these two groups of men, the extreme differences noted in Table 4 become blurred. For example, only marijuana, poppers and cocaine are used by more than 5% of the gay sample more than once a week, while more than 5% of the heterosexual men use marijuana and cocaine. Gay men as a whole are more likely than their heterosexual neighbors to use poppers, barbiturates and amphetamines on at least a weekly basis, while the heterosexual men were more likely to use cocaine that often. It seems that while gay men use a wide variety of drugs, they are not given, in general, to the frequent use of specific drugs.

These findings regarding the prevalence of drug and alcohol use among gay men, it should be added, stand in considerable contrast to those generated through the use of opportunistic sampling techniques. Previously, through the use of such techniques, it has consistently been found that very high rates of problem drinking and problematic drug use exist within this community. It may be that the use of opportunistic sampling techniques within gay male populations causes an inordinate proportion of bar patrons to be sampled, a group long known to contain a disproportionate number of problem drinkers. Conversely, by using household sampling techniques, transient, institutionalized or homeless gay men are not interviewed, thereby presumably lowering estimated rates of problem drinking within this population. Nothing in these findings should be used to indicate that the current concerns within the gay community regarding drug and alcohol problems are misplaced. That is, even the "conservative" estimates described here would indicate that drug and alcohol consumption patterns may well be a threat to a significant proportion of the gay male community.

From these data the conclusion can be drawn that efforts to prevent the spread of further HIV infection within this community which rely on preventing drug or alcohol use are probably inappropriate. This conclusion is derived from the finding that gay men as a group do not drink much differently than heterosexual men, and as a group are moderate drinkers. Attempts to prevent alcohol consumption within the entire community would thus likely be perceived as punishing the group for the excesses of a few. Further, since enormous effort has been made to forbid drug use to this and all other communities and yet the relationship between drug use and unsafe sex remains, suggests that efforts to prevent further HIV infection by restricting the access to drugs or alcohol within this community are doomed to failure.

PREVENTION OF HIV INFECTION RELATED TO DRUG AND ALCOHOL USE

Before prevention implications are discussed, a short summary of the most important findings outlined in this paper will be given. A strong association exists between the use of drugs and/or alcohol during sexual activity and the kinds of sexual activity implicated in HIV infection. Although interpretation of the meaning of this association is difficult, the consequences of HIV infection are so profound that health education and prevention strategies which are designed to prevent further HIV infection related to drug and alcohol consumption during sexual activity seem warranted. Although health education strategies are important in lowering rates of further HIV infection, further prevention efforts might also be justified.

Prevention efforts aimed at HIV infection related to drug and alcohol consumption during sexual activity could logically be designed to prevent drug and alcohol consumption among gay men or to prevent the *combination* of drug and alcohol use and sexual activity. Alcohol consumption patterns among gay men do not appear to be appreciably different from heterosexual men and are, for the most part, rather moderate. Further, prevalence of use of a long list of drugs over a six month period is considerably higher among gay men than among heterosexual men. Since draconian measures have been used to prevent drug use among Americans, the argument that prevention of the use of drugs or of alcohol will result in lower rates of participation in risky sexual behavior is not tenable. That is, it is difficult to imagine what further measures could be adopted to prevent drug use, and efforts to prevent alcohol use are not likely to be any more effective, especially within a social group for which bars serve a function as community centers. Thus, strategies to prevent HIV infection related to drug and/or alcohol use would best be aimed at preventing the *combination* of substance use and sexual activity.

Several strategies seem appropriate in devising a prevention campaign designed to prevent further HIV infection related to the combination of sexual activity and substance use. First, a set of specialized health education campaigns should be attempted. The gay male population should be informed of the excess risk for HIV infection associated with combining drug and/or alcohol use with sexual activity. Secondly, the social characteristics of those who are most likely to combine these two activities should be identified: specialized health education campaigns should be attempted for these subgroups of the gay male population.

A primary assertion of this paper, however, has been that effective prevention policy must rely on more than health education strategies. Strategies must also be devised which change the social ecology of risk for HIV infection. For example, as previously mentioned, since there is a strong association between high risk sexual activity and substance use, prevention workers could work with the management of gay bars to change the setting in which a considerable proportion of new sexual contacts within the gay male world are made. That is, condoms could be made readily available in gay bars so that if sex occurs between bar patrons, it occurs after being in a setting where condoms are easily obtained. Secondly, bartenders could be given training in "server intervention" strategies, so that bartenders are shown how to identify and "cut off" inebriated patrons.[13] This second suggestion has the added advantage of lessening the chances for bar owners' liability should patrons become involved in a serious driving accident after drinking in their bar.

A third possibility for decreasing drug and alcohol-related HIV infection has to do with changing the social environment in which risk occurs. Given that a strong association exists between substance use and high risk sexual behavior, it is difficult to imagine a more dangerous setting for gay men to meet socially and to make new sexual contacts than one which must market a drug to survive economically. Yet, social environments for gay men to meet are nearly entirely limited to bars, and this is especially the case outside of major urban settings. Thus, it may be that finding settings for gay men to meet socially which are not located in a setting which encourages alcohol use might well work to minimize new HIV infection.

Prevention efforts to control the AIDS epidemic are at present our main hope in containing the toll taken by this epidemic. Although there have been impressive decreases in high risk behaviors for exposure to HIV among gay men as a group, a proportion of this population has continued to risk infection throughout the course of this epidemic. For this, and other reasons, it can be determined that while health education is a necessary cause for changing behaviors on a population-wide basis, it is not a sufficient cause. Prevention policy must utilize techniques which, in combination with health education efforts, serve to further decrease behaviors which are implicated in HIV transmission. It is argued that these prevention techniques will work best if they are designed with an understanding of the conditions under which individuals decline to comply with risk-reduction guidelines. It appears that the use of drugs and alcohol during sexual contact is one such condition. Three prevention strategies are suggested, which in addition to health education efforts, might be used to minimize HIV transmission related to the use of drugs and alcohol during sexual activity. It is hoped that the approach suggested in this paper might also be adopted to develop prevention strategies for other populations at risk of HIV infection, e.g., those who share needles or those who have sexual contact with needle users.

NOTES

1. McKusick, L., J. Wiley, T. Coates, et al., Reported changes in the sexual behavior of gay men at risk for AIDS. San Francisco, 1982-1984: The AIDS Behavioral Research Project, Pub Hl Rep 1986: 100:622-629.

2. Weber, J., T. Coates and L. McKusick. Denial may hinder AIDS risk reduction: The AIDS Behavioral Research Project. 1986. Manuscript under review.

3. McKusick, L., W. Horstman and T. Coates, AIDS and the sexual behavior reported by gay men in San Francisco. 1985: Am J Pub Hl 75:493-496.

4. Stall, R., L. McKusick, J. Wiley, et al., Alcohol and drug use during sexual activity and compliance with safe sex guidelines for AIDS: The AIDS Behavioral Research Project. 1986. Health Ed Quart 13:359-371.

5. McKusick, L., J. Wiley, T. Coates and S. Morin, Predictors of AIDS behavioral risk reduction: The AIDS Behavioral Research Project. Presented at

the New Zealand AIDS Foundation Prevention Education Planning Workshop, November, 1986.

6. Ostrow, D., Barriers to the acknowledgement of the role of drugs and alcohol in AIDS and the treatment of drug and alcohol problems in persons with AIDS and HIV infection. Paper presented at the Joint Meeting of the American Medical Society on Alcoholism and Other Drug Dependencies and the Research Society on Alcoholism, San Francisco: April, 1986.

7. Fifield, L., T. Latham and C. Phillips, Alcoholism in the Gay Community: The Price of Alienation, Isolation and Oppression. A Project of the Gay Community Services Center, Los Angeles, CA. 1977.

8. Lohrenz, L, L. Connelly, L. Coyne, K. Spare, Alcohol problems in several midwestern homosexual communities. J Stud Alcohol. 1978, 39:1959-1963.

9. Morales, E. and M.A. Graves, Substance Abuse: Patterns and Barriers to Treatment for Gay Men and Lesbians in San Francisco. San Francisco Community Substance Abuse Services. San Francisco: Prevention Resource Center, 1983.

10. Nardi, P., Alcoholism and homosexuality: a theoretical perspective. 1982. J Homosex 7:9-25.

11. Wiley, J., W. Windelstein, T. Piazza, et al., The San Francisco Men's Health Study: Recruitment of the Cohort. Survey Research Center Working Paper #49; Berkeley, CA: Survey Research Group, University of California, Berkeley, August, 1986.

12. Cahalan, D., R. Roizen and R. Room, Alcohol problems and their prevention: public attitudes in California. In: R. Room and S. Sheffield, eds. *The Prevention of Alcohol Problems: Report of a Conference*, pps. 354-403. Office of Alcoholism, Health and Welfare Agency, Sacramento, California, 1974.

13. Jernigan, D. and J. Mosher, Preventing alcohol-related high risk sex among gay men. Paper delivered at the 114th annual meeting of the American Public Health Association. Las Vegas, 1986.

Heterosexual Contacts of Intravenous Drug Abusers: Implications for the Next Spread of the AIDS Epidemic

Debra L. Murphy, PhD, MPH

SUMMARY. There is a scarcity of knowledge about the sexual behavior patterns of the intravenous drug abuser (IVDA), despite the potential role of this group in the heterosexual transmission of AIDS. Using a representative sample of 93 clients from the Addiction Research and Treatment Corporation (ARTC), who consented to be anonymously interviewed, this study investigated sexual behavior patterns and practices of these intravenous drug abusers. Over half of the 758 sexual contacts reported by this sample were non-IVDAs. Results indicated that IVDA males in comparison with IVDA females reported a significantly greater percentage of heterosexual contacts which were non-IVDAs. However, when controlling for needle sharing, this association was significant only for those who shared. The findings indicate the importance of targeting, not only IVDAs with regards to health educational interventions, but also sexually active non-IVDAs, especially females, in communities where intravenous drug abuse is prevalent.

The author is affiliated with the Columbia University School of Public Health Socio-medical Sciences Division. Reprint requests should be addressed to Debra Murphy, PhD, MPH, Columbia University School of Public Health Socio-medical Sciences Division, 600 West 168th St., NY, NY 10032.

Data from this study were presented by the author at the III International Conference on AIDS, the 49th Annual Scientific Meeting of the C.P.D.D., and the 95th Annual Meeting of the Amer. Psych. Assoc. Convention. The author conducted this research while at the Addiction Research and Treatment Corp., and the author gratefully acknowledges ARTC and the cooperation of the participants of this study.

INTRODUCTION

Intravenous drug abusers (IVDAs) are at second highest risk for the contraction of AIDS and are potentially the bridge of infectivity to the non-intravenous drug abusing heterosexual community.[1] To date, sexual contact with IVDAs who have AIDS is the predominant mode of transmission for U.S. born non-intravenous drug abusing heterosexual cases of AIDS.[2]

Much has been reported with regards to the at-risk for AIDS needle sharing practices of IVDAs; however, knowledge about the sexual behavior patterns and practices of the IVDA is scarce. Given the potential for IVDAs as the major mechanism for the epidemic spread of the AIDS virus to the heterosexual community, more information with regards to sexual behavior patterns of IVDAs is imperative. This is crucial in order to understand the future epidemiological spread of AIDS so that interventions can target those groups most at risk for heterosexual exposure via IVDAs. The purpose of the following study was to examine the sexual contact patterns of IVDAs within and outside of the intravenous drug abusing group and how this may affect the epidemiological spread of the HIV virus.

METHOD

Subjects

The sample consisted of randomly selected clients currently enrolled in 4 Addiction Research and Treatment Corporation (ARTC) Methadone Maintenance Clinics, who volunteered to be anonymous participants in the study.

The sample size was 93: 58% male, 42% female, 54% Black, 35% Hispanic, and 11% White with a mean age of 35. These demographics for the sample of 93 and the total ARTC population of 2100 were statistically indistinct.

Procedure

During medication hours, August 31, 1986 through October 31, 1986, trained interviewers asked randomly selected clients for their anonymous participation in the study. If the client consented, the interviewer accompanied the participant to a private room where the client was engaged in a one hour structured interview with regards to drug use and sexual behavior patterns and practices during the period 1977-1985.

Analysis

The data were analyzed using the Chi-Square test to determine probability of chance occurrence.

RESULTS

Twenty-one individuals refused to participate and 14 consented, but were not interviewed, because of recurring scheduling difficulties. They were 77% Black, 20% Hispanic, 3% White, 63% male, and 37% female with a mean age of 37. Among these clients as compared with the study group, the sex and age distributions were statistically indistinct; however, the number of Blacks was significantly greater and the number of Hispanics and Whites was significantly less.

Of those sampled, 88 (95%) reported heterosexual contacts between the period 1977 to 1985. One male and two females indicated bisexual activity and homosexual behavior was reported by one male and one female.*

The majority of the sample reported having only one sex partner between 1977 and 1985 (See Table 1). Of those who were heterosexual, females were significantly more likely than males to have had a single sex partner (61% versus 25%). Twenty-eight percent of

*Reflected in the remaining analyses are participants indicating heterosexuality including prostitution. Analyses not shown which included and excluded non-heterosexuality and or prostitution indicated no difference in the findings presented in this article.

Table 1

Number of

Reported Contacts by Sexual Orientation

	1	2-3	4-11	>12	N
Heterosexual	%	%	%	%	
Males	25	21	26	28	53
Females	61	17	17	5	35
Bisexual					
Males			100		1
Females		50	50		2
Homosexual					
Males	100				1
Females	100				1
Total	40	20	23	17	93

the males were in sexual contact with greater than 12 persons; however, only one (5%) female indicated this number of partners and she admitted to prostitution. Three of the five nonheterosexuals reported one or two sexual contacts. The remainder reported greater than four partners.

Next, the sexual contacts by this sample which were within and outside of the intravenous drug abusing group were investigated. This sample of drug abusers reported sexual contact with a total of 758 persons between 1977 and 1985. Of these contacts, 498 (66%) were reported as being non-intravenous drug users in comparison with 260 (34%) which were drug injectors (p < .001). The females in the sample indicated a total of 105 sexual contacts. Of these, 42 (40%) were noninjectors. In contrast, the males in the sample reported a total of 653 sexual contacts, and 457 (70%) were individuals who did not inject drugs. Of the remainder, 196 (30%) of the 653 sexual contacts by males were IVDAs, compared with 63 (60%) of the 105 sexual contacts by females, who were drug injectors. These results suggest that male in comparison with female IVDAs were significantly more likely to report sexual contacts who were non-IVDAs than IVDAs (See Table 2).

Next, how needle sharing affected this association of gender with intravenous drug abusing status of partner was examined. Twenty-two percent of the sample reported ever sharing needles between 1977 and 1985. There was no significant difference between females and males (24% vs. 21%). Controlling for needle sharing, the association of gender with intravenous drug abuse status of sexual contacts was no longer significant for those who did not share needles (See Table 3). Among those who shared, the percentage of sexual contacts who were non-IVDAs was significantly greater for males than females (78% versus 22%). Whether or not males reported a significantly greater percentage of non-IVDA sexual contacts than females was very much dependent upon whether the clients shared needles.

Among females, sharing needles was highly significantly associated with sexual contact with injecting partners. Table 4 shows that the percentage of IVDA partners reported by females who shared was more than twice that reported by females who did not share (78% versus 32%). In contrast, among males who shared, the per-

Table 2

Reported Number of IVDA and Non-IVDA Partners

By Sex

	Non-IVDA (N=498) %	IVDA (N=260) %
Male	70	30
Female	40	60
Total	66	34

*
Chi-Square=35.65, p<.0001

Table 3

Reported Number of

IVDA and Non-IVDA Sex Partners

By Sex Controlling for Needle Sharing

	Did Not Share		Shared	
	Non-IVDA	IVDA	Non-IVDA	IVDA
	(N=392)	(N=183)	(N=106)	(N=77)
	%	%	%	%
Males	68	32	78	22
Females	68	32	22	78

*
p=N.S.   *
                              Chi-Square=69, <.0001

centage of IVDA sexual affiliations was significantly less than that for those who did not share (22% versus 32%).

DISCUSSION

The finding that over half of this sample reported contact with non-intravenous drug abusers, provides strong support that IVDAs may play a very important role in exposing the HIV virus to the general non-IVDA community. The finding that males, in comparison with females, reported a greater percentage of non-IV drug abusing sexual contacts was similarly reported by Des Jarlais,[3] who suggested that IVDA males have to be having sexual relationships with non-IVDA females. This was based on a study of heroin abusers entering treatment in New York City in 1983 in which the male-to-female ratio was 2.70:1 and a survey of 50 male IVDAs who reported that 40 of their 53 heterosexual partners were non-IVDAs.

However, in the present study, when controlling for needle sharing, this association appeared significant only for those who shared. These findings suggest that, of those who share needles, male IV-

Table 4

Reported Number of

IVDA and Non-IVDA Sex Partners

By Sharing Controlling for Sex

	<u>Male</u>		<u>Female</u>	
	Non-IVDA	IVDA	Non-IVDA	IVDA
	(N=456)	(N=197)	(N=42)	N=63)
	%	%	%	%
No Share	68	32	68	32
Share	78	22	22	78
Total	70	30	40	60

*
Chi-Square=4, p<.05

*
Chi-Square=24,p<.0001

DAs may pose a greater likelihood of exposing the virus to individuals outside of the intravenous drug abusing circle; especially to females in the general population who may not presently belong to high risk groups.

In the Des Jarlais[3] study, the contacts were regular partners and did not include prostitutional contacts. Similarly, the findings from the present study may be limited to non-prostitutional partners, since only one person in the sample admitted to prostitution. Females who prostitute to support their drug addiction have been suggested as a significant possible source of spread of the HIV virus.

It is possible that those who refused to participate in the study were more likely to have been prostitutes and therefore not represented in this sample. However, there is other evidence which suggests that prostitution may not be very prevalent within the ARTC population. For instance, other studies at ARTC have indicated low numbers of partners reported by clients.[4] Laboratory indices routinely collected on all clients reflect minuscule numbers of females who have contracted sexually transmittable diseases.[5] Similarly, in the present study, only 3% of the clients reported having had either syphilis, gonorrhea, or herpes between 1977 and 1985.

The association of females who share needles with IV-drug using partners, in contrast to those who do not, is consistent with research which has indicated that the use of needles has certain social, psychological, and sexual connotations associated with it which may enhance the likelihood that those who share will affiliate with others who do similarly.[6] Individuals who share needles may be more intimately involved in the drug subculture than those who do not. Consequently, research which has indicated a greater propensity for females, in contrast to males who are IVDAs, to affiliate sexually within the IV-drug abusing circle, because female drug users are initiated into drug use by males, may be more applicable for those who share than those who do not.

Other research has additionally indicated that females also receive monetary and emotional reinforcement for continued drug dependency from male partners, whom female intravenous drug abusers identify as their major suppliers. However, while drug use among females may be initiated and maintained by males, the reverse situation rarely occurs.[7]

The results from the present study indicate that there is considerable sexual affiliation of IVDAs with partners outside of the intravenous drug abusing group. Males who share needles and are at high risk may be especially likely to spread the virus to females who do not presently belong to high risk groups. The importance of targeting, not only, IVDAs, with regards to health educational interventions, but also, sexually active non-IVDAs, especially females, in communities where intravenous drug abuse is prevalent cannot be over-emphasized.

NOTES

1. Drucker, E. AIDS: The eleventh year. New York State Journal of Medicine, May, 1987: 87:5:255-256.

2. Chambers, M., Centers for Disease Control, Personal Communication, June, 1987.

3. Des Jarlais, D., Wish, E., Friedman, S., Stoneburner, R., Yancovitz, S., Mildvan, D., El-Sadr, W., Brady, E., Cuadrado, M., Intravenous drug use and

the heterosexual transmission of the human immunodeficiency virus. New York State Journal of Medicine, May 1987: 87:5: 283-286.

4. Brown, L.S., Evans, R., Murphy, D., Primm, B., Drug use patterns: implications for the acquired immunodeficiency syndrome. Journal of the National Medical Association, 1986:78:1145-1151.

5. Unpublished data, Addiction Research and Treatment Corporation, 1987.

6. Howard, J., Borges, P., Needle sharing in the Haight. Journal of Health and Social Behavior, 1970:11:3:220-230.

7. Prather, J. Women's use of licit and illicit drugs. In J. Lowinson and P. Ruiz (Eds.), Substance Abuse: Clinical Problems and Perspectives. Baltimore: Will. and Will., 1981:729-738.

Sterile Needles and the Epidemic of Acquired Immunodeficiency Syndrome: Issues for Drug Abuse Treatment and Public Health

Peter A. Selwyn

SUMMARY. The debate over the provision of sterile injection equipment to intravenous drug users, as a means of preventing the spread of the AIDS epidemic, has a number of political, ethical, and clinical implications. The issue has in some respects been inappropriately dichotomized as a conflict between public health agendas and the traditional priorities of drug treatment. The relevant issues include: (1) the existence of evidence for needle-sharing as a route of transmission of human immunodeficiency virus among intravenous drug users; (2) the role of needle scarcity as a factor promoting needle-sharing behavior, and evidence for the ability of drug users to change such behavior; (3) the possibility of increased needle availability leading to increased prevalence of intravenous drug abuse; (4) the possibility that the provision of sterile needles would compromise treatment efforts among drug abusers currently or potentially engaged in the treatment system. These issues are discussed in light of relevant existing data; a multilevel strategy for AIDS prevention among drug users is suggested, addressing both the availability of sterile injection equipment and the promotion of drug treatment goals.

Peter A. Selwyn, MD, Department of Epidemiology and Social Medicine, Montefiore Medical Center, 111 East 210th Street, Bronx, NY 10467.

INTRODUCTION

The issue of whether to facilitate access of intravenous drug users (IVDUs) to sterile needles and syringes, as a means of reducing the transmission of human immunodeficiency virus (HIV), has a number of political, ethical, and clinical implications for society at large and particularly for substance abuse treatment professionals. Since early in the current epidemic of the acquired immunodeficiency syndrome (AIDS), debate has continued on the potential risks and benefits of such a policy as a strategy for AIDS prevention. A potentially destructive dichotomy has arisen between what could be described as a public health intervention model, which would seek to halt the transmission of HIV through the free distribution of needles and syringes, and a drug treatment model, which would reject anything short of an abstinence-oriented approach as an unacceptable incentive to drug abuse.

The principal questions involved may be summarized as follows: (1) Is there evidence that the sharing of contaminated or nonsterile needles is the route of HIV transmission among IVDUs? (2) Is there evidence that IVDUs have changed behavior and/or are capable of changing behavior in response to the AIDS epidemic, and that scarcity of injection equipment is a key factor in promoting needle sharing? (3) Will increased availability of sterile needle/syringes lead to increased levels of IV drug use or accidental needlesticks among the general population? (4) Will increased availability of sterile needles compromise treatment of IVDUs already or potentially engaged in the treatment system, by making continued IV drug use more attractive? These questions will be examined sequentially in the discussion which follows.

NEEDLE-SHARING AND THE TRANSMISSION OF HIV

Several cross-sectional and case-control studies of risk factors for HIV infection among IVDUs have indicated that use of nonsterile needles and visits to "shooting galleries" are associated with HIV infection as indicated by the presence of serum HIV antibody, and that the use of sterile needles may be protective.[1-3] In a 1984 study of

308 IVDUs recruited from methadone maintenance and drug detoxification programs in New York City, in which 51% of subjects were found to be HIV-seropositive, HIV antibody was associated with the frequency of drug injection and the percentage of injections in shooting galleries.[1] Traditional needle-cleaning practices, such as washing injection equipment with water or alcohol, were not found to be protective against HIV infection. In a 1985 study of 497 patients in a methadone maintenance program in the Bronx, 35% of subjects were determined to be HIV-seropositive. In this study, seropositivity was associated both on univariate analysis and logistic regression analysis with several variables reflecting the extent and frequency of nonsterile needle use and the number of visits to shooting galleries since 1978.[2] A recent study of 281 heterosexual IVDUs in San Francisco, while reporting a lower seroprevalence level of 10%, found that seropositivity was significantly associated with needle-sharing; the odds ratio for seropositivity among those who shared needles with two or more persons vs. those who injected alone was 5.4.[3] The data from these and additional studies provide consistent and compelling evidence that needle-sharing and the use of nonsterile injection equipment are important routes of HIV transmission among IVDUs.

BEHAVIOR CHANGE AMONG INTRAVENOUS DRUG USERS IN RESPONSE TO AIDS

Studies in New York City suggest that an increased street demand for sterile needles has recently developed, with drug dealers adopting practices such as selling two needles at a time (in case one should become obstructed with blood), repackaging used needles to appear new, and asking higher prices for sterile equipment.[4] Several surveys of knowledge and attitudes about AIDS among IVDUs have indicated widespread recognition of needle-sharing as an AIDS risk factor, with concern about contracting the disease frequently expressed by study subjects.[4-7] New York City data from studies conducted in 1984 and 1985 among patients in methadone programs and narcotic-addicted inmates at a major detention facility indicate that up to 60% of respondents reported needle-use behavior changes in response to concerns about AIDS, ranging from total abstinence to cessation or decrease in needle-sharing while still injecting drugs.[6,7]

Despite the changes in behavior noted among IVDUs in these studies, the data indicate that a certain proportion of IVDUs continue to engage in unsafe injection practices, regardless of the awareness of risk. In one study, the explanation most frequently given for continued needle-sharing despite knowledge of the risk of AIDS was characterized as the "need to inject drugs, with no clean needle available," offered by 46% of persistent needle-sharers.[6] Other common explanations, particularly among detention center inmates, were "only share with close friend or relative," or "enjoy social aspects of needle-sharing." These findings indicate that the scarcity of sterile needles may indeed be an important factor in needle-sharing among certain subgroups of IVDUs, whereas among others the social context of needle-sharing may be more important. The data further suggest that the provision of sterile equipment, if contemplated, must be linked with expanded educational programs and other behavioral interventions addressing the social networks of needle-sharing among IVDUs. A recent study of peer organizations of IVDUs in the United States and Europe suggests that peer pressure and the collective self-organization of drug users may help reinforce safer needle use practices and eventually encourage abstinence, similar to the process in which self-organization has promoted safer sex practices among homosexual men.[8]

NEEDLE AVAILABILITY
AND THE PREVALENCE OF DRUG ABUSE

Since 1984, the city of Amsterdam in the Netherlands has conducted a needle-exchange program whereby addicts may exchange used needles and syringes for new sets at designated sites throughout the city. This program was initiated as a joint effort between municipal and health authorities and a local association of drug users ("Junkies' Union"), and was part of an expanded effort to provide drug treatment and widespread AIDS education. In 1985, over 100,000 needles and syringes were provided, and surveillance data indicate that there have been no signs of an increase in needlestick accidents among the general population nor of any increase in measures of the prevalence of IV drug use in that city.[9] Data are not yet available regarding the impact of this intervention on the spread of HIV infection in Amsterdam, but will soon be forthcoming. Several pilot needle-exchange programs are now being contemplated in different areas in the United States.

EFFECTS OF NEEDLE AVAILABILITY ON TREATMENT GOALS

Firstly, one must reject the argument that fear of AIDS should be an appropriate and intentional societal disincentive for the use of IV drugs. There are legal precedents to discourage strategies seeking to deter proscribed activities by increasing their medical hazards, in the case of contraceptive sales to minors. In addition, there is little empirical evidence that such strategies would even be likely to have the desired effect. In a 1977 decision, the U.S. Supreme Court unanimously rejected an argument by New York State that minors' sexual activity would be deterred by a prohibition on buying contraceptives, stating that, "It would be plainly unreasonable to assume that (the State) has prescribed pregnancy and the birth of an unwanted child (or the physical and psychological dangers of abortion) as punishment for fornication. We remain reluctant to attribute any such 'scheme of values' to the State."[10] Similarly, one must not implicitly or explicitly seek to prescribe AIDS as punishment for IV drug use. While abstinence is clearly a desirable treatment goal for all IVDUs, there is ample evidence that this is not always possible in the short-term, and that for certain subgroups of chronic abusers, IV drug use continues in spite of the best efforts of treatment professionals. Particularly for these high-risk subgroups, one might argue that even while treatment efforts continue, one should take steps to ensure that if needle use persists it be done more safely. Indeed, especially for active users not engaged in the treatment system, programs to encourage safer needle use might in fact be an effective first step in the attempt to involve drug users in treatment, in settings where abstinence-oriented messages may not be well received. The experience with peer organizations of IVDUs, as alluded to above, supports the potential utility of this approach.[8]

CONCLUSIONS

While the policy debate over the provision of sterile needles continues, public health interventions have already been initiated to address nonsterile needle use as a route of HIV transmission. Educational materials and community outreach efforts providing instruction on safe techniques for needle-cleaning have already been implemented in several urban areas in the United States. Proposals for needle exchange programs are also now being considered in New York City and New Jersey, both for IVDUs on waiting lists for treatment and among high-risk users already in treatment settings but continuing to use needles. Such programs will use short-term outcome measures such as retention in treatment, urine toxicologies, clinical evaluation, and the assay of returned used needle/syringe sets for different ABO blood types as markers of the effectiveness of needle-exchange strategies. Long-term measures of the utility of such programs would clearly need to include rates of HIV-seroconversion among participants as compared with rates among nonparticipants.

Without expanded opportunities for treatment and widespread educational efforts, the provision of sterile needles may be rightly criticized as little more than a cynical and desperate attempt to halt the AIDS epidemic. However, given the evidence that needle scarcity may promote needle-sharing in certain populations of IVDUs, to continue to restrict access may ultimately prove to be equally harmful. Arguments that the continued restricted access to sterile injection equipment will serve as a disincentive to intravenous drug abuse have little evidence to support them, and have questionable ethical implications. The data from the Netherlands which indicate no measurable increase in intravenous abuse following the implementation of a needle-exchange program provide further evidence that such proposals need not be summarily dismissed for fear of encouragement of drug use. Whatever strategies are developed must be controlled, evaluable, and sensitive both to the urgency of the AIDS crisis and the underlying commitment to treating drug abuse effectively.

The experience with self-organization of drug users suggests that peer support networks may be particularly useful in the effort to promote behavior change in this population. Education about needle hygiene and the provision of sterile needles may be necessary as one facet of this effort if behavior change is to be perceived as a credible and feasible alternative among the large group of IVDUs not currently engaged in the treatment system. At this critical historical moment, it would appear that the burden of proof would be on those who advocate abstinence as the single message for AIDS education among drug users to demonstrate why other strategies addressing needle use and availability should not be developed as part of a multi-level response to the AIDS epidemic. The risk of not formulating such strategies, because of a perceived conflict with the traditional goals of drug treatment, may be to make the drug treatment community increasingly isolated both from drug users themselves and from the emerging public health policy debate over AIDS prevention.

NOTES

1. Marmor M, Des Jarlais DC, Cohen H, et al. Risk factors for infection with human immunodeficiency virus among intravenous drug abusers in New York City. AIDS 1987; 1:39-44.

2. Schoenbaum EE, Selwyn PA, Feiner C, et al. Prevalence of and risk factors associated with HTLV-III/LAV antibody among intravenous drug abusers in a methadone program in NYC. Presentation at International Conference on AIDS, Paris, France, June 23-25, 1986.

3. Chaisson RE, Moss AR, Oriski R, et al. Human immunodeficiency virus infection in heterosexual intravenous drug users in San Francisco. Am J Public Health 1987; 77:169-172.

4. Des Jarlais DC, Friedman SR, Hopkins W Risk Reduction for AIDS among intravenous drug users. Ann Intern Med 1985; 103:755-759.

5. Ginzburg HM,: Intravenous drug users and AIDS. Pub Health Rep 1984; 99:206-212.

6. Selwyn PA, Cox CC, Feiner C, et al. Knowledge about AIDS and high-risk behavior among intravenous drug users in New York City. Presentation at International Conference on AIDS, Paris, France, June 23-25, 1986.

7. Friedman SR, Des Jarlais DC, Sotheran JL AIDS health education for intravenous drug users. Health Education Quarterly 1986; 13:383-393.

8. Friedman SR, Des Jarlais DC, Sotheran JL, et al. AIDS and self-organization among intravenous drug users. Int J Addictions 1987; 22:201-219.

9. Buning EC, Coutinho RA, et al. Preventing AIDS in drug addicts in Amsterdam (letter). Lancet 1986; i:1435.

10. The Committee on Medicine and Law. Legalization of non-prescription sale of hypodermic needles: a response to the AIDS crisis. Record of the Association of the Bar of the City New York 1986; 41:809-820.

Neurocognitive Impairment in Alcoholics: Review and Comparison with Cognitive Impairment Due to AIDS

Melvin I. Pohl, MD

SUMMARY. This review indicates that cognitive dysfunction and dementia, specifically caused by chronic alcohol use, consist of subtle changes in cognition which are easily missed and/or forgotten. The nature of the dysfunction and specific neuropsychiatric and neurodiagnostic tests will be reviewed. There are important treatment implications regarding these subtle changes in mental function. Treatment implications will also be summarized. Finally, a comparison will be made between the cognitive impairment experience by patients with Acquired Immune Deficiency Syndrome from that experienced by alcoholics.

INTRODUCTION

It has been well established that alcohol toxicity causes brain dysfunction. Many of these changes will be described in the following paper. Excessive use of alcohol and drugs has an effect on the immune system. There is an extremely high incidence of Acquired Immunodeficiency Syndrome in intravenous drug users. There are over 50,000 cases of Acquired Immunodeficiency Syndrome in the

Melvin I. Pohl, MD, is Medical Director, Pride Institute, Eden Prairie, Minnesota and Clinical Instructor in Family Medicine, University of Nevada School of Medicine, Department of Family and Community Medicine. For reprint requests write to: Melvin I. Pohl, 2810 West Charleston Boulevard, F54, Las Vegas, NV 89102.

United States today. Perhaps 2 million more are infected with and capable of transmitting AIDS through intimate sexual contact and needle sharing behaviors. Human Immunodeficiency Virus (HIV) cause AIDS and as part of that syndrome causes cognitive impairment.

This paper will describe the cognitive impairment which occurs in alcoholics. It then will compare and contrast cognitive impairment that occurs in people with AIDS from that occurring in alcoholics. There is much overlap in these areas making differentiation extremely difficult, if not impossible.

ALCOHOLIC BRAIN DYSFUNCTION

Cognitive impairment in alcoholics has been shown to be present in 70-90% of patients after detoxification.[1,2,3] Patients in most of these studies had been drinking for an average of ten years. The impairment was in cognition; that is, the complex and dynamic central integration function of the brain. Cognitive function is responsible for manipulation of information and switching of ideas. Impaired cognition results in the loss of the fine edge of intellect. Cognition is responsible for learning (systematized ability to recall encoded information), reasoning, planning, judging, organizing, forming concepts, solving problems (thinking logically and testing hypotheses), and self-evaluating.

Severity of alcoholic brain dysfunction ranges from subtle cognitive impairment to severe organic brain syndrome subdivided into alcoholic dementia and amnestic syndrome or Korsakoff's psychosis.

Alcoholic dementia is defined as (1) Loss of intellectual abilities resulting in impaired social and occupational function; (2) Memory deficit; (3) Impaired cortical function (as described above, including abstracting capacity, judgment, cognition) or a personality change; (4) Clear consciousness.[4] With alcoholic dementia, patients seem older. They have a slower ability to process and have a prolonged reaction time. They also have slower access to memory stores and a failure to form key associations. Language and intelligence are intact, which often make this diagnosis difficult.

With progressive cognitive impairment, there is progressive deterioration of global cognitive function over time rather than sequential development of specific deficits.[4] There is extreme variability between alcoholic dementia as it occurs from patient to patient, regardless of consumption (time or frequency).

Etiology

Etiologies of alcoholic dementia include direct chemical toxicity of brain tissue as well as other factors. Chemical toxicity results in a decrease in glial cells as well as a decrease in dendritic arborization. There may be inhibition of brain protein synthesis, thus impairing neuronal function as well as altering endorphin levels. Other etiological factors to be considered include preexisting impairment such as inherited and environmental state prior to the onset of drinking; trauma to the central nervous system secondary to falls, fights and accidents; toxicity of alcohol to the organism resulting in metabolic acidosis, liver disease, and nutritional deficiencies; and affective disorders, especially depression.[5]

Diagnosis

Cognitive dysfunction is demonstrated most effectively by a series of neuropsychiatric tests and corroborated by neuropathological changes as defined by radiologic scanning techniques of the brain. When evaluating an alcoholic who appears demented it may be wise to perform neurocognitive testing to define the nature and level of impairment. Therapeutically it may be helpful to reevaluate this patient at six month intervals to document improvement, if any. If level of consciousness is disturbed in an alcoholic, it is imperative to at least get a CAT scan of the brain to rule out intracranial bleeding. If dementia persists, a more thorough medical work-up is in order.

Intelligence Tests

Firstly, intelligence tests, typically the Wechsler Adult Intelligence Scale (WAIS), are practically normal after the detoxification period.[5] There are abnormalities frequently found on block design (a time pressured test of visual perception, analysis of forms, and ability to reproduce); object assembly (ability to form a whole from parts in a time pressure fashion); and digit symbol (also a time measured test which calls for matching numbers with symbols). Intelligence tests essentially demonstrate defects in performance corresponding with decreased visual organization, slower reaction time, and decreased ability to handle complexity.

Neuropsychological Test Battery

The most frequently used neuropsychological testing battery is the Halstead-Reitan neuropsychological test battery.[5] These tests assess behavior deficits associated with organic dysfunction. There is some improvement over the course of four weeks suggesting reversible toxicity of alcohol as well as alcoholic liver disease, metabolic acidosis, nutritional deficiencies, etc. The most commonly abnormal tests in these batteries are described in Table 1. Alcoholics

TABLE 1

TEST	METHOD	FUNCTION TESTED
I. CATEGORY TEST	1. Involves four graphics of varying shapes and color.	1. The ability to utilize differential hypotheses. 2. Requires verbal and central integrative skills. 3. Involves ability to abstract both visually and spatially.
II. TACTUAL PERFORMANCE TEST (TPT)	1. Utilizes a wooden form board, timed and done right handed, left handed, and with both hands. 2. With the patient blindfolded, they must fit certain forms into the board.	1. Measures psychomotor coordination. 2. Involves strategy, efficiency and the ability to organize. 3. Demonstrates the ability to process and encode spatial relationships. 4. Abnormalities in the TPT indicate right hemisphere dysfunction, especially the parietal lobe.
III. TRAILS B	1. Requires connecting letters with numbers. 2. Performed in sequence and must switch mental sets from numbers to letters.	1. Evaluates perceptual motor integration and speed.
IV. TAPPING TEST	1. Involves tapping with the index finger.	1. Evaluates fluidity, speed and coordination.

invariably evidence dysfunction on all four of the tests in Table 1, indicating impairment of higher brain function. They tend to be impulsive and arbitrary, particularly on the TPT, and become frustrated quite easily. There is evidence of impairment in the ability to abstract concepts and to perform functions requiring perceptual motor integrity. There is also impairment of ability to perform judgment tasks and immediate adaptive functions.[7]

Imaging Techniques

Computerized axial tomography (CAT) scans confirm cortical atrophy with prominent sulci, ventricular enlargement, increased cerebral CSF, and questionable cerebellar atrophy, which is confirmed by magnetic resonance imaging. Serial CAT scans performed over four weeks indicate a decrease in cerebral CSF and an increase in cerebral density with abstinence.[8,9] Positron emission tomography (PET) indicate decreased rates of glucose metabolism in cortical areas, thalamic nuclei, and basal ganglia. This correlates with decreased long term memory tasks with specific decrease in cell densities noted in the dorsomedial nucleus of the thalamus.[10] EEG's show increased beta activity and decreased alpha activity corresponding to these abnormalities.

Prognosis

Recovery of neurocognitive impairment occurs and is progressive over time. With advancing age (greater than 40 years old), this recovery is slower and less complete. Even in the youngest of alcoholics, total recovery may not occur. One to three weeks after the last drink, there is significant improvement in short term memory, visual, and visual-spatial function. After one to two months, there is an increased retention and increased ability to analyze complex visual stimuli as evaluated by the category test. There is, however, some impairment in visual-spatial function and attention after as long as five years. Recovery is felt to relate to regrowth of neurons, collateral sprouting, and alterations in neurotransmitter synthesis and release, as well as altered post synaptic sensitivity.[11]

Treatment

Treatment implications, in view of the above information, are significant. When working with people who are cognitively impaired, one must first be aware of this impairment. One must go quite slowly, giving the patient time to learn and master new materials. Repetition and reinforcement creatively done to avoid boredom and fatigue are essential for adequate learning. Cognitive impairment may interfere with the ability to understand and accept the disease concept and the principle of abstinence (the first step in Alcoholics Anonymous). It is important to differentiate these difficulties as being cognitively mediated rather than subconsciously related to denial, resistance, or decreased motivation. Because of the decreased ability to learn, it is important that the patient be able to develop cues to apply across situations and learn appropriate problem solving strategies in principle and then in specific situations.[12]

Because of the extent of cognitive impairment, these patients need education which is straightforward, clearcut and filled with examples. For each bit of information that they learn, they need firm and consistent reinforcement both in the short term and in the long term. Aftercare is an absolutely critical component of recovery in the cognitively impaired alcoholic. Because of their cognitive difficulties, support systems become all the more important, consistent with sponsorship and fellowship groups.

It is apparent that even with all of the above measures carefully followed there may come a point when patients who feel and look normal, who've been clean and sober for several weeks to several months, may still be unable to process, judge, organize, plan, and self-evaluate. It is here that the concept of spirituality and faith in a higher power would seem most applicable for the alcoholic patient. This may in part explain why for successful recovery through traditional 12-step programs an intrapsychic change needs to take place. What better vehicle to use than a "spiritual" connection with a "power greater than oneself" to combat the sense of unexplainable powerlessness over a mind that *seems* to be normal, which is not able to function properly to solve problems and formulate concepts. It may be that this is the reason why unlike any other illness, alcoholism and drug dependency seem to depend upon a spiritual link for a positive outcome.

COGNITIVE IMPAIRMENT AND THE
ACQUIRED IMMUNODEFICIENCY SYNDROME (AIDS)

When differentiating cognitive impairment caused by alcoholism from that of AIDS, there are several factors to consider. First, the etiologies for both are multifactorial. In chemical dependency, there is primarily toxic damage, and in AIDS there is predominantly infectious damage. In both, affective disorder may play a role in the development of cognitive impairment. It is critical to assess for affective disorders since they are treatable (with or without medications). The incidence of impairment is high: as high as 80% in both groups, and is sorely underestimated in both groups. With alcoholism, impairment involves learning ability, abstracting skills, judgement, and processing functions. With alcoholism, the dysfunction most often is mild, while in AIDS the dysfunction eventually becomes severe. In alcoholism, the dysfunction stops with abstinence and becomes worse if the drug is continued.

With alcoholic cognitive impairment there is atrophy with increased ventricular size. With AIDS, there is a generalized slowing which is related to subcortical dysfunction. On scanning, there is similar atrophy and cellular changes with increased ventricular size. Microscopically, there is vacuolation of cells, which may be evident on MRI scanning.[13]

In AIDS, the progression is usually relentless. There are few satisfactory treatment modalities for AIDS dementia though AZT does penetrate the blood brain barrier and may have a positive effect on mental function. With alcoholic dysfunction, the changes are gradually and mostly all reversible.

Changes with AIDS are irreversible. Treatment implications for both include becoming and remaining aware of the dysfunction and not forgetting it during treatment.

In evaluating the patient with AIDS dementia, cerebral dysfunction may be caused by a variety of disturbances and frequently is related to metabolic, infectious, or malignant consequences of the underlying immune deficiency. Most specifically, disorders of electrolytes, magnesium, pH, pO_2, blood glucose, liver function, and renal function must be evaluated in each AIDS patient. Subsequently, if the patient is exhibiting cerebral signs, opportunistic infections must be ruled out. Common opportunistic infections of the brain include toxoplasmosis gondii, Cryptococcal meningitis, cytomegalovirus cerebritis, and herpes simplex encephalitis. Malignancies frequently affect the brain with AIDS, particularly Kaposi's sarcoma and even more frequently lymphoma. CAT scanning and MRI's may be helpful in differentiating etiologies. If a space-occupying lesion is present, brain biopsy may become necessary to differentiate treatable from untreatable etiologies for neurological deficits.[14] In the absence of provable infection or space-occupying lesion in an AIDS patient, dementia must be assumed to be caused by actual HIV cerebritis or meningitis.[15,16]

It is important not to expect more of these patients than they are capable of. With the alcoholic, it is important to delay as much treatment requiring cognitive function as possible since they will improve with time of abstinence. With the AIDS patients, it is important to do as much cognitive treatment as possible (will planning, treatment decisions, etc.) as soon as possible since the impairment will often get worse. For the alcoholic, their cognitive impairment and knowledge about it is a source of frustration and upset. They feel they should be better, and do not take solace in the fact that their brain is impaired. For the patient with AIDS, and particularly for the care-giver of that patient, information about impairment may be helpful, resulting in relief. As people with alcoholism require sponsors, patients with AIDS, because of cognitive impairment, do well with buddies, which is the basis of many successful AIDS service organizations' programs. Firm and loving reinforcement as well as a dependence on spiritual factors is a key technique for alcoholics and may well prove to be essential for the treatment of people with AIDS.

As discussed earlier, there are alcoholic patients who develop AIDS. These patients may be suffering from two types of dementia resulting in profound impairment from which they are unlikely to recover even with abstinence from alcohol. In an alcoholic who remains cognitively impaired despite a period of abstinence, it is appropriate to proceed with neuropsychological evaluation. This evaluation must include ruling out HIV disease with antibody testing. In appropriate patients who have antibodies to HIV, it is appropriate to order T cell subsets and Helper/Suppressor ratios. If HIV disease is present, it then becomes important to evaluate the patient for the possibility of underlying opportunistic infection (such as toxoplasmosis, cytomegalovirus or herpes) or malignancy (such as lymphoma or Kaposi's sarcoma).

CONCLUSION

With the increasing number of cases of AIDS, it will soon be impossible for a clinician in medical practice to not come in contact with AIDS. For those treating alcoholics, it becomes more important to familiarize themselves with the Acquired Immunodeficiency Syndrome, particularly with its effects on the central nervous system. It is important to be alerted to the signs and symptoms of cerebral dysfunction in all alcoholic patients with particular attention being paid to those patients who do not clear with prolonged abstinence. Finally, in treating all alcoholics, it is extremely important to remember the patients may be more limited than they appear. These limitations will have an impact on successful treatment and need to be taken into consideration as the patient recovers.

NOTES

1. Claiborn JM, Greene RL. Neuropsychological Changes in Recovering Men Alcoholics. J Studies on Alcohol 1981; 42:757-765.

2. Grant I, Adams K, Reed R. Normal Neuropsychological Abilities of Alcoholic Men in Their Late Thirties. Am J Psychiatry 1979; 136:1263-1269.

3. Long JA, McLachlan JFC. Abstract Reasoning and Perceptual-Motor Efficiency in Alcoholics: Impairment and Reversibility. Quart J Stud Alc 1974; 35:1220-1229.

4. Miller WR, Orr J. Nature and Sequence of Neuropsychological Deficits in Alcoholics. J Studies on Alcohol 1980; 41:325-337.

5. Echardt MJ, Martin PR. Clinical Assessment of Cognition in Alcoholism. Alcoholism: Clinical and Experimental Research 1986; 10:123-126.

6. Tarter RE, Edwards KL. Multifactorial Etiology of Neuropsychological Impairment in Alcoholics. Alcoholism: Clinical and Experimental Research 1986; 10:128-134.

7. Etkoff, Louis. Personal Communication. 1987.

8. Carlen, PL, Penn RD, Fornazzari L, Bennett J, Wilkinson DA, Phil D, Wortzman G. Computerized Tomographic Scan Assessment of Alcoholic Brain Damage and Its Potential Reversibility. Alcoholism: Clinical and Experimental Research 1986; 10:226-232.

9. Chao HM, Foudin L. Symposium on Imaging Research in Alcoholism, Introduction to the Symposium. Alcoholism: Clinical and Experimental Research 1986; 10:223-224.

10. Johnson JL, Adinoff B, Bisserbe JC, Martin PR, Rio D, Rohrbaugh JW, Zubovic E, Echkardt M. Assessment of Alcoholism-Related Organic Brain Syndromes with Positron Emission Tomography. Alcoholism: Clinical and Experimental Research 1986; 10:237-240.

11. Goldman MS. Neuropsychological Recovery in Alcoholics: Endogenous and Exogenous Process. Alcoholism: Clinical and Experimental Research 1986; 10:136-144.

12. McCrady BS, Smith DE. Implications of Cognitive Impairment for the Treatment of Alcoholism. Alcoholism: Clinical and Experimental Research 1986; 10:145-150.

13. DeLa Monte S, Moore T, Hedley-Whyte E. Vacuolar Encephalopathy of AIDS. NEJM 1986; 315:1549-1550.

The Response of State Agencies to AIDS, Addiction, and Alcoholism

Chauncey L. Veatch III, JD

SUMMARY. State agencies have been profoundly impacted by the AIDS epidemic. In the absence of a vaccine that would prevent AIDS or of medicines that would cure it, the primary strategies of such agencies have focused on reducing the spread of AIDS by promoting cessation of high risk behaviors and thus preventing or slowing its transmission. Recent research indicates that the primary route of AIDS transmission into the general heterosexual population is by intravenous (IV) drug abusers, who directly account for about 17% of AIDS cases nationwide. Reducing the spread of AIDS within this group would not only reduce the overall toll of the disease but should limit its spread to the population at large. Infection by the human immunodeficiency virus (HIV) can be minimized by reducing or eliminating certain high-risk activities. In the IV drug using community, the primary intervention strategies include: educating IV drug users about the hazards of AIDS and sharing of needles; enrolling them in treatment programs to reduce drug use; promoting the use of new or sterilized syringes and needles among those who will not abstain from drug use; and discouraging high-risk sexual activity among those who are already infected by HIV. The State of California has already increased the number of treatment slots for IV drug users and, through the Department of Alcohol and Drug Pro-

---

Chauncey L. Veatch III, JD, is Director, California Department of Alcohol and Drug Programs, 111 Capital Mall, Sacramento, CA 95814.

The author would like to acknowledge the assistance of Kurt Klemencic, AIDS Coordinator, California Department of Alcohol and Drug Programs (ADP); and consultation provided by M. Douglas Anglin, PhD, and Joseph Arnold, MS, ADP's Drug Abuse Information and Monitoring Project (UCLA). The editorial review of Thomas Maugh, science writer for the Los Angeles Times, is appreciated.

grams, is scaling up its educational, prevention, and intervention activities, particularly those related to safe sex, promoting the cessation of IV drug use, and improving equipment hygiene by those who continue use.

OVERVIEW

It is hard to believe that it has been only six years since acquired immune deficiency syndrome, or AIDS, was first identified. During a relatively short period, the disease progressed rapidly from affecting only isolated members of groups that some perceived as deviant, to a full-fledged threat to public health. Intravenous, intramuscular, and subdermal drug users, collectively known as parenteral drug users, comprise the second largest population group at risk for AIDS in the United States. As of February 1, 1988, the U.S. Centers for Disease Control (CDC) reported over 50,000 cumulative AIDS cases. Of that total, 17% (5,099) were parenteral drug users and an additional 8% (2,318) were homosexual or bisexual parenteral drug users.

Scientists have moved determinedly to characterize the AIDS syndrome, to isolate the punitive agent, human immunodeficiency virus (HIV), and to develop reliable tests that identify infected individuals. Their efforts have shown how the disease is spread and have given public health professionals insight into the natural history of the disease. Hence, we have been able to mount effective intervention and reduce the spread of the disease.

Virtually all researchers who have studied AIDS and its transmission have reached one conclusion about controlling its spread: those individuals who engage in high-risk activities MUST change their behavior to avoid becoming infected or to avoid infecting others.[1,2,3]

The unprecedented nature of the disease, the sex-linked and drug-linked nature of transmission, the long incubation period, and the variety in its presentations made it difficult in the early years for researchers and concerned agencies to achieve an understanding of the disease and its implications. Hence uniform and appropriate intervention strategies were delayed in development and implementation. As a result, both the public at large and the public health community have criticized federal and state agencies for what has been perceived as slowness in implementing policies to restrict the spread of AIDS. Moreover, discussion of public health strategies such as case tracing or premarital testing have provoked strong reactions by some members of the affected groups, even though these approaches have been commonly accepted for other diseases. The issue involved here is an important one: the preservation of individual privacy for affected individuals weighed against public health concerns.

Only after the nature of the epidemic was more fully understood did such a diversity of national agencies as the CDC, Public Health Service (PHS), Food and Drug Administration (FDA), National Institutes of Health (NIH), and Social Security Administration (SSA) become heavily involved. They now realize the impact this disease could have on their constituencies because it affects individuals of all ages, sexes, races, places of residence, and sexual orientations who engaged in "high-risk" activities, who have had a blood transfusion, or who were born to an HIV-exposed mother.

Against this backdrop, the role of government agencies in the crisis is particularly sensitive because it places demands upon public administrative bodies to make decisions of the most fundamental and controversial nature. Especially difficult are decisions in which the rights of the individual must be balanced against the needs of society. Such decisions are normally the province of legislative or judicial bodies.

Many state agency directors, including myself, are extremely sympathetic to the problems of those individuals whose personal behaviors may place them at risk, and it is imperative that a general consensus is achieved regarding policies that respect these individuals' rights. At the same time, however, we must recognize that transmission of AIDS into the society at large would produce widespread tragic results.[4] All necessary resources must therefore be directed toward halting the spread of HIV infection by any and all methods that can be scientifically and ethically justified.

The experiences of New York and New Jersey have made it clear that parenteral drug users are the single most important conduit for transmitting HIV to non-drug-using heterosexuals.[5,6,7] Since no vaccine or medical treatments currently exist for AIDS, our greatest hope for preventing an epidemic spread of the infection in all populations lies in limiting viral transmission. Parenteral drug users are particularly at risk, and thus are the focus of this paper. First, however, let me briefly review the dimensions of the AIDS problem.

DIMENSIONS OF THE AIDS PROBLEM

According to the CDC, more than half of the estimated 1 to 1.5 million individuals already infected with HIV will eventually develop AIDS.[7] The latency period between HIV infection and development of overt AIDS averages at least 4 years in adults. Therefore, most of those individuals who will develop AIDS by 1991 have already been infected with HIV.[8]

Based on an empirical model that uses reported cases of AIDS,[9] the cumulative number of cases of AIDS (meeting the current CDC definition) will exceed 270,000 by 1991 (see Table 1). During that year, more than 145,000 AIDS cases will require medical care and an additional 54,000 will die, bringing the cumulative number of AIDS-related deaths to more than 179,000. We must remember, however, that some cases may go unreported, so that the empirical model may underestimate the number of cases and deaths.

The proportion of AIDS cases occurring in the heterosexual nonparenteral drug using population is predicted to rise from the current level of 7% of the total to more than 9% in 1991, or nearly 7,000 new cases among heterosexuals in that year alone. In addition, there will be more than 3,000 cumulative cases of infant and pediatric AIDS by the end of 1991.

The geographic distribution of AIDS is also changing. Currently, more than 40% of AIDS cases are concentrated in New York City and San Francisco. By 1991, however, those two cities are expected to have less than 20% of the total cases, and a much larger number of cities and states will have to confront the epidemic.[2]

As the number of AIDS cases increases and the distribution of these cases diffuses beyond the homosexual/bisexual male population, existing service organizations (many of which are volunteer

TABLE 1: DIMENSIONS OF THE AIDS PROBLEM

NUMBER OF:	1986	1991[a] ESTIMATES
UNITED STATES		
HIV seropositive individuals	1-1.5 million	9-13.5 million
cases meeting current CDC criteria (cumulative)	30,396*	270,000
cases needing hospital care	14,000[b]	145,000
cumulative deaths	17,542*	179,000
CALIFORNIA		
cases meeting current CDC criteria (cumulative)	6795**	45,000
cumulative deaths	3316**	31,800

a) Confronting AIDS[9]
b) Green[10]
*as of 2/10/87
**as of 12/31/86

organizations supported by the gay community) will be unable to handle the new demand for their services.

Two reasons for this are: (1) the service organizations are already reaching their maximum capacity; and (2) the services provided by these organizations are now oriented to the homosexual community and may not meet the different needs of other affected populations. This shortage will then create an additional strain on hospitals and other health care facilities, which will dramatically increase costs to the health care system as a whole.

For example, one study projects that the care of AIDS patients in 1991 will require 4.4% of all hospital beds in New York City, compared to only 1.62% now (Table 2). The situation in San Francisco

TABLE 2: IMPACT ON HOSPITAL SYSTEMS

AREA/REGION	BEDS AVAILABLE	%FOR AIDS 1986	%FOR AIDS 1991
New York	39,832	1.62	4.40
San Francisco	6,806	2.06	9.53
Other U.S. Cities	1,076,102	0.14	0.96
Total U. S.	1,122,740	0.20	1.14
Total cost	$766 million		>$4.3 billion

Source: Green, Singer, & Wintfield[10]

will be even worse: AIDS care will require 9.53% of all hospital beds. compared to 2.06% now.[10] According to the U.S. Public Health Service (PHS), the total national cost for AIDS care in 1991 may lie between $8 billion and $16 billion,* compared to only $766 million now.[11] These estimates do not include the cost of care for individuals with AIDS-related complex (ARC) or the costs associated with experimental therapies or lengthened survival times, so it underestimates the direct costs of HIV infection in 1991.

PARENTERAL DRUG USERS

Since the advent of the AIDS epidemic, it has become very important to determine the extent of parenteral drug abuse and to identify those involved—not only for the traditional reasons of education, prevention, and treatment, but also because of the larger public health considerations.[6] It is necessary to expend every effort

*Green, Singer, and Wintfield[10] estimate 1991 hospital costs alone at > $4.3 billion.

to save the lives of as many drug abusers as possible, and to limit the spread of HIV infection into the non-drug-using community.

Historically, little has been known about the demography, natural history, and sociology of the parenteral-drug-abusing population. We know that an estimated 750,000 Americans inject illicit drugs intravenously, intramuscularly, or subdermally at least once a week; a similar number inject drugs less frequently. We also know that heroin users comprise the large majority of the parenteral drug user population, but we do not know precisely how large that proportion is.[8] Finally, we know that heroin users are concentrated in large cities, but we need to know with greater certainty what percentages of users are found in these areas and the distribution between urban and rural areas.

We have even less information about parenteral users of other substances. The heroin addict is not alone in being at risk for exposure to HIV. Also at risk is the occasional drug user who shares a needle or syringe when self-administering cocaine or amphetamines. Data from national samples of drug abuse treatment programs indicate that 80% of all clients seeking treatment, regardless of the drug of abuse at time of treatment, have injected drugs intravenously during the previous year.[5] This population is a particular problem because individuals are highly motivated to stay hidden from view. Thus, the population with which state and city agencies must deal is largely a matter of speculation.

Another problem for assessment efforts is the great heterogeneity of parenteral drug users. Many disparate groups are represented: the wealthy and/or highly educated, including executives, entertainers, and sports figures; blue collar workers many of whom maintain steady jobs during periods of active drug use; and, of course, those who are disenfranchised from the mainstream of society in our ethnic barrios, ghettos, and other pockets of poverty. Drug abuse is also found among individuals of all sexual persuasions. To address the cultural, ethnic, and sexual life-styles of these very different groups requires diverse but coordinated efforts by state agencies.

The actual incidence of AIDS and the prevalence of HIV seropositivity varies widely among parenteral drug users throughout the country and the world (Tables 3 and 4). In New York City and

TABLE 3: CLINICAL AIDS AND IV DRUG USE

SAMPLE	N	%
All AIDS CASES[+]	30,632	100.0
IV drug history	7,460	24.3
IV drug use (sole risk)	5,123	16.7
LOCATION OF IV AIDS CASES	4,204	100.0
New York[1]	2,530	60.2
New Jersey[2]	907	21.6
Florida[3]	301	7.2
California[4]	155	3.7

AIDS CASES IN SELECTED POPULATIONS	N	IV History N	%
Prisoners[*]	766	728	95
Adult women[+]	2,267	1,056	47
Children[*]	273	148	54

Sources:
[+] CDC MMWR Mar. 23, 1987
[1] New York State Department of Health, January 28, 1987.
[2] New Jersey State Department of Health, April 1, 1987.
[3] Florida State Department of Health, March 2, 1987.
[4] California State Department of Health, Dec. 31, 1986.
*New York Times, April 26, 1986

Newark, for example, the prevalence of HIV infection is estimated at over 50%. In San Francisco, by contrast, the prevalence among parenteral drug users is estimated at about 16%. The remainder of California is, fortunately, very early in the epidemic among parenteral drug users. Evidence from a relatively few studies indicates that the prevalence of HIV seropositivity outside San Francisco is only about 2%.[9] Aside from the figures for New York and Newark, these numbers can be viewed with some optimism because they suggest that immediate intervention could severely limit the number of parenteral drug users who will contract AIDS.

Transmission of HIV to the heterosexual population seems certain to continue. Infected bisexual men and parenteral drug users of

TABLE 4: SEROPREVALENCE OF AIDS VIRUS ANTIBODIES IN IV DRUG USERS

AREA/SAMPLE	YEAR	POSITIVE (%)
UNITED STATES		
New York (Des Jarlais)	1986	50-64
New Jersey (Rutledge)	1986	50-60
San Francisco (Moss, Chaisson)	1986	10-16
Los Angeles (Mascola)	1986	2
EUROPE		
Italy (nationwide*)	1987	37
Edinburgh (Robertson)	1986	51
Spain (Rodrigo)	1985	37
Greece (Papaevangelou)	1985	2

*Twenty of different regions were sampled and seroprevalence ranged from less than 10% in two regions to greater than 50% in two regions. A total of 8928 IVDU attending assistance centers were screened. Social Medicine Department of Ministry of Health and AIDS National Task Force, Rome Italy, 3/6/87.

both sexes will transmit HIV to the broader heterosexual population, where it will continue to spread as a result of high-risk sexual activities. Transmission to adolescents may represent a particular problem because they are among the least susceptible to public education campaigns and the most likely to carry out high-risk sexual activities.[9] If intervention efforts can prevent the spread of AIDS within the parenteral-drug-using population and can curtail or limit the number of unsafe sexual contacts between parenteral drug users and the community at large, then we can greatly reduce the escalation of the AIDS epidemic.

Although their contribution to the spread of AIDS is not as direct as parenteral drug use, the role of alcohol and nonparenteral drug consumption in eliciting high-risk behaviors may be an important one. A recent study concluded that gay men who consumed alcohol during sexual episodes were more than twice as likely to engage in high-risk sexual practices as men who did not drink.[12] Additionally, the study indicated that consumption of other drugs during sexual activity increased the propensity for engaging in high-risk behaviors. While illicit drugs (e.g., marijuana and "other drugs") were shown to produce a greater effect than legal and more easily accessible drugs (e.g., alcohol, "poppers"), all were associated with some increase in high-risk activity. The study also showed a correlation between the frequency of drug use during sex and the incidence of specific high-risk sexual practices that were likely to transmit HIV. Although the significance of these reports is not yet fully explicated, these correlations may continue to play a significant role in HIV infection.

This study did not examine alcohol or drug use as a disinhibiting factor in the period *prior* to episodes of sexual activity; it seems likely, however, that such use preceding the initiation of sex could, for some drugs, have an even larger effect than concurrent use because there is a greater time period during which the drugs or alcohol can take effect. The potential importance of such prior use has been recognized by both the National Institute on Drug Abuse (NIDA) and the National Institute of Allergy and Infectious Diseases (NIAID), and further research is being solicited.

The California Legislature has also addressed drug use as a possible cofactor affecting the progression from HIV infection to the development of clinical AIDS. California law requires establishments that sell poppers to post a sign warning that inhalation of alkyl nitrates may be harmful to health, may affect the immune system, and has been associated with the development of Kaposi's sarcoma, an AIDS-related cancer.[13]

EXPERIENCES OF THE GAY COMMUNITY

Historians tell us that we can often learn from mistakes made by our predecessors. This learning is useful for politicians, academicians, businessmen, and military leaders alike. In the case of the AIDS epidemic, where millions of lives may be at stake, those of us directing social policy must be aware of the experiences of the gay community with AIDS.

As early as 1981, members of the gay and medical communities were concerned about the potential for spread of the then-unidentified virus that was destroying the immune systems of hitherto healthy young gay men. Evidence slowly accumulated that this virus would continue to spread, not only among homosexual or bisexual men, but also in IV drug abusing, hemophiliac, and other populations.[14]

Local, state, and federal agencies were trying to adjust to the emergence of a new disease that gave few indications about its potential scope or seriousness. Because there was no existing infrastructure to respond to the rapidly increasing needs of AIDS patients, the gay community, as the most affected group, rallied together to provide assistance. This support took many different forms.

For example, community organizations in Los Angeles, such as Aid for AIDS and AIDS Project Los Angeles, raised significant amounts of money from donations for direct patient care.[15] Other funds have been raised to provide education about preventive measures for individuals at high risk for HIV transmission; again, these funds came largely through contributions by members of the gay community.[16]

One effective educational approach taken by the gay community was to conduct outreach activities by training volunteers at various community service centers about AIDS and safe-sex behaviors.[17] These volunteers disseminated the information to individuals in the community. This effort, in conjunction with media campaigns about the AIDS epidemic, proved to be effective not only in making AIDS a household word, but also in getting the gay community to shift toward safer sexual practices.[16]

Leaders within the gay community also helped organize a good secondary-care system, much of it voluntary. This system has helped to keep health care costs down by offering the AIDS patient home health care provided by community service centers and local AIDS projects. Such support enables AIDS patients to be discharged from the hospital sooner, thereby reducing their cost of care. One reason that New York City's cost of care is so high is that there are few hospices for those in the terminal phases of the disease. Where hospices, whose costs are $55-$65 per day, or other less expensive facilities for AIDS patients are not available, care must be provided by hospitals, whose costs average $963 per day.[17,18]

Hospice services are also preferred modes of care on humanitarian grounds. Hospice care provides the AIDS patient a more caring, more dignified, and more comfortable, environment in which to spend the terminal stages of the disease; in such an environment, the individual is permitted to retain some degree of decision-making ability.

Not all of the experiences of the gay community can be translated to the parenteral-drug-using community. There are no comparable visible and active leaders, and probably never will be, because drug users do not have the same sense of community; neither are the same levels of economic support available. For the same reasons, supportive facilities are not likely to be organized by users themselves. It will thus fall to national, state and local agencies to provide both leadership and facilities.

BEHAVIORAL CHANGE IN THE
DRUG ABUSING POPULATION

The ideal solution to the problem of AIDS in the drug-using population would be to convince drug users to stop using drugs. Unfortunately, the high frequency at which parenteral drug users return to some level of drug injection after treatment,[1,19] suggests that this is an unlikely possibility. Because the discovery of an effective vaccine against HIV or a cure for AIDS is still years away, it is necessary for public health officials to convince drug users to modify their behavior so that they are less likely to contract or spread the disease.

An outline for attacking the spread of AIDS within the parenteral-drug-using population is found in the guidelines and recommendations offered by the Coolfont Conference,[8] NIDA,[20] the National Academy of Sciences Institute of Medicine,[9] and the World Health Organization.[21]

Those recommendations that directly affect AIDS, addiction, and alcoholism focus on: development of an information base to monitor the spread of HIV exposure within the parenteral-drug-using population; voluntary testing for HIV infection with strict adherence to confidentiality laws; increasing the number of slots at drug treatment and methadone maintenance centers; development of diversified and innovative educational campaigns aimed at changing unsafe sexual and needle-sharing behaviors; trial distribution of bleach, condoms, and sterile syringes; and increased funding for research (Table 5).

A recurrent suggestion for the first step in combating the AIDS epidemic among parenteral drug users is the extensive testing of blood samples to identify carriers of HIV. The identification of HIV carriers by race, ethnicity, age, sex, geographic area, and sexual preference allows us to predict future, costly demands on the health care system and to target our educational efforts.[9]

Because of ethical and legal restraints, we can now conduct such testing only on individuals who request HIV antibody testing. Thus, the individual must be motivated to participate in the screening.

TABLE 5: RECOMMENDATIONS MADE BY COOLFONT, WHO, NIDA, NAS

---

* Reduction of IV drug use and prevent HIV infection among drug users

* Development of innovative educational campaigns to encourage modification of high-risk behavior of IV drug users and their sexual partners

* Encouragement of voluntary HIV testing (possibly within drug abuse treatment programs)

* Increasing treatment availability

* Identification of co-factors to AIDS in drug users

* Development of a national commision on AIDS

* Increasing national expenditures for AIDS education and prevention

---

That has not been a large problem with members of the gay population, who are motivated both by the fear of their own death and the fear of infecting those with whom they share intimate sexual experiences.

Similar levels of cooperation are not generally expected from members of the drug-using population. They are much more familiar with the prospect of death because the very activity in which they participate exposes them to numerous opportunities for death, including overdose, hepatitis B, endocarditis, and violent confrontations with drug traffickers or with police.[22]

Fear of a lingering, painful illness may be a more effective motivator, and it has been suggested that parenteral drug users be shown video tapes that demonstrate the cruel realities of enduring HIV infection.[23]

HIV testing is especially important where pediatric AIDS is a consideration, such as for sexual partners of drug users who anticipate pregnancy, those who already are pregnant, and newborn infants. Special efforts should be directed at these groups. Our goal in the abstract thus seems simple enough: just find the best ways to produce the desired behavioral change in parenteral drug users and their sexual partners, and then implement programs to act on that knowledge.

EDUCATION AND PREVENTION

Experience has shown that the parenteral risk group, and those at risk due to relationships with parenteral drug abusers, do not necessarily respond well to the types of appeals that have been effective in lowering the rate of HIV transmission among members of the gay community.[24,25] Unfortunately, only recently has research been conducted to determine the most effective methods for motivating these groups to modify their behavior.

In part, the lack of research may arise from the misconception among some in the scientific community that parenteral drug abusers are a "system-resistant population" who do not respond well to any social intervention efforts. That belief is weakened substantially by several recent surveys which indicate that many drug abusers are concerned about AIDS and are willing to listen to reason.[25,26]

Local efforts to educate drug abusers, in fact, have been strikingly effective where intense outreach and community education techniques have been employed, most notably in San Francisco and Newark.[1]

Nonetheless, increasing receptivity among the targeted audience, whether gay or parenteral drug users, necessitates adapting the educational message and medium to insure not only the education of the group but also the implementation of behavioral changes. It has been shown that behavioral change campaigns that simply broadcast information via television or newspaper advertising do not necessarily change behavior, even among such receptive audiences as the gay community. Furthermore, the hesitancy to use specific and direct language in describing drug hygiene and sexual behavior may be reducing the effectiveness of this approach.[23]

Drug users not in treatment are more difficult to reach because they are less accessible and they are generally less educated and less organized than the gay community. Communication among street drug users is generally oral rather than written or printed because many lack literary skills or lack the motivation to read; in some instances, they lack trust in most written materials.[26] As a group, furthermore, drug users have little in common other than their participation in illicit activities. Apart from the drug distribution system that fulfills their physical needs and the formal treatment system that addresses their clinical needs, there are few organizational or institutional groups to unify individuals in this high-risk category.[27]

There is one specific legal issue that is seen by many as impeding efforts to halt the spread of AIDS through the parenteral drug using community. That issue involves the restrictions on the sale or distribution of sterile syringes and needles in many states, including California. Increasing their legal availability has been recommended by some public health officials. However, many other health officials, as well as law enforcement officials, are opposed because they believe the relaxation of such restrictions could lead to increased parenteral drug use.[9]

In 1986, public health officials in Amsterdam began distributing sterile syringes to parenteral drug users because they felt that any effort to reduce HIV transmission overrode all other concerns. Preliminary and anecdotal evidence indicate that these efforts do not, in their present form, reduce the spread of seropositivity.[28] For these reasons efforts at this time to decrease the legal restrictions on distribution of needles and syringes in the United States are likely to fail.[29] We must therefore urge parenteral drug users not to share equipment. If sharing persists, they must be taught how to sterilize their equipment. These preventive measures must also be combined with graphic and explicit educational campaigns that will penetrate the recalcitrant drug users' inherent authority resistance.

Despite the difficulties associated with altering the behavior of drug abusers, carefully designed information delivery strategies about the increasing scope of the AIDS epidemic has prompted many of them to attempt to reduce their risk of HIV exposure by reducing drug use or by practicing safer injection behaviors.[25] Local outreach programs may have proven to be the most effective way to reach the parenteral-drug-using population. New programs might be modeled after the successful San Francisco Mid-City Consortium to Combat AIDS program, where vials of bleach are distributed along with instructions on how to sterilize "works" before they are shared.[30]

In New York City, ex-addicts employed by drug treatment programs have begun to emulate educational programs conducted in the gay community. For example, the Association for Drug Abuse Prevention and Treatment provides services for parenteral drug users who have AIDS. Its members also educate past, present, and potential drug users about the risk of HIV transmission.[25] Such one-on-one sessions are generally considered to be the most effective method of reaching drug abusers.

Counseling should probably also be extended into the criminal justice system. Statistics from New York City demonstrate that as many as 85% of individuals booked for a criminal violation have used drugs recently.[27] One approach might be to test all arrestees for drug use and provide counseling about the dangers of AIDS to all who test positive. A similar program of testing and counseling might prove useful for other populations, such as prostitutes and street hustlers. The cost of such counseling, currently estimated to be about $500 per counselee,[29] is considerably cheaper than the $60,000 + cost of treating an AIDS patient.

Physicians must play a crucial role in the successful education of both drug abusers and their sexual and needle-sharing partners. If a drug user seeks medical treatment, it is imperative to capitalize on the opportunity to provide direct information about AIDS exposure and ways to prevent transmission. As some physicians may be uncomfortable about AIDS or drug users,[31,32] AIDS information must become part of our medical school curricula, the continuing education programs, and the training of other health care professionals.

Additionally, we should not overlook the role that religious organizations could play in AIDS prevention programs. While a number of controversial issues surround the participation of various church-affiliated organizations and officials in AIDS suppression efforts, the potential outreach capacity of the various denominations may be crucial in an overall strategy of AIDS containment.

To further promote educational efforts, any agencies dealing with drug users—who thus have expertise in the area—must appropriately inform and educate the population at large (including other relevant agencies, social policy planners, and legislators) to elicit a coordinated response to existing problems. Such cooperation is essential for the implementation of effective programs. It is unfortunate that the stigma of drug abuse, particularly parenteral drug abuse, reduces these groups' incentive to understand the issues surrounding drug abuse and the drug abuser.

The advent of AIDS in this high-risk group has substantially aggravated existing stereotypes and prejudices and has made rational approaches to dealing with the substantive issues much more difficult. The public and political controversy surrounding pragmatic efforts at intervention can be considerable. Unfortunately, this controversy can delay or dilute our efforts significantly.

Finally, we need to sharply reduce the lag time that elapses after agencies have acquired new information about AIDS and before they have disseminated it to the appropriate groups within their constituency. This slowness of response creates the mistaken impression among the affected individuals and their advocacy groups that the agencies are reluctant to help them. It is clear that state agencies must take steps to reduce this inherent bureaucratic inertia. Ideally, this should be accomplished by streamlining existing infrastructures rather than by creating new ones—which would only add to the current lag time.

One approach might be to centralize AIDS efforts wherever possible and to increase coordination among state agencies and their federal, local, and community counterparts. The California Department of Health Services and the Department of Alcohol and Drug Program have recently increased coordination efforts by the exchange of personnel who act as liaison between the agencies.

DRUG TREATMENT AND AIDS

The one measure most strongly recommended by researchers and health agencies to reduce HIV transmission associated with drug use is an immediate increase in the number of treatment slots at drug treatment centers.[20,21,22,25] Because of the current shortage of such slots, drug users who wish to quit frequently must wait weeks or months to be enrolled. Two problems result from protracted waiting periods: (1) The individual remains at risk for HIV infection because intervention is delayed; and (2) loss of motivation may occur, so that the applicant is no longer interested in treatment when a slot becomes available.

Furthermore, from a purely economic point of view increasing the number of treatment slots can reduce overall costs to society. The total cost of treating an AIDS patient can range from $50,000 to $150,000 per case, whereas the cost of drug abuse treatment can be as little as $3,000 per year in an outpatient program.[9]

California has responded to the growing need for the treatment of parenteral drug abuse by easing the requirements for admission to treatment programs and increasing the number of slots.

In 1986 and 1987, the number of slots for methadone detoxification treatment increased by 21% to 4,282, while the number of slots for maintenance has increased by 19% to 12,601 (Table 6). Furthermore, as of January 1987, licenses were pending for 9 new clinical programs for drug abuse treatment. In the past two years, in contrast, only four new clinics were licensed.

Additional measures taken by the State Department of Alcohol and Drug Programs (ADP) include pending regulations that would allow methadone maintenance treatment facilities to increase the

TABLE 6: CHANGES IN TREATMENT SERVICES IN CALIFORNIA

# OF TREATMENT SLOTS	1985	1987	% INCREASE
DETOX	3,540	4,282	21
METHADONE	10,590	12,601	19

OTHER CHANGES	1986	1987
EMERGENCY CAPACITY	10% max over authorized capacity	limited to # of staff & facility size
ADMISSION CRITERIA	2 years of IV drug use	1 year of IV drug use
	2 prior failures	no prior treatment failures
READMISSION CRITERIA	28 day waiting period	7 day waiting period
AIDS COORDINATOR ON SITE	no	mandatory

number of emergency treatment slots from the normal 10% of authorized capacity to a ceiling dependent upon number of staff and facility capabilities. Criteria for emergency admission to these programs will change from a two-year history of parenteral drug use and two prior treatment failures to one year of parenteral drug use and no prior treatment failures. The waiting period for readmission will be reduced from 28 days to the federally mandated length of seven days. Another important component of these regulations requires that each treatment facility have an AIDS coordinator on site to encourage HIV antibody testing, and that each provide education and counseling services for both staff and clients. Twenty-one existing clinical programs have applied for changes under these new regulations.

In addition to providing a larger number of treatment slots, we must also provide improved training for the staff of existing and new treatment centers. Many of the staff are poorly informed about AIDS and apprehensive about their own contacts with potential HIV carriers. It will be very difficult to counsel drug abusers about AIDS if treatment center staff are not properly educated and trained.

Providing treatment for drug users will assure a reduction in opportunities for needle-sharing activities. But, an effective educational campaign that inspired drug users to seek treatment for their addiction in order to avoid exposure to HIV could overtax existing facilities. We must make provision for such increased demand, because allowing a motivated addict to go untreated during an epidemic such as we now face with AIDS is tragic.

The primary objective of state agency officials must be to reduce the transmission of HIV. To ensure public acceptance, the most palatable approach to that goal is an increase in the treatment services for parenteral drug users. Such an approach addresses two different diseases, addiction and AIDS, simultaneously and it does not give the appearance that either an agency or the public is condoning drug use.

ISSUES FOR CONSIDERATION

To recapitulate, then, we have a number of needs that must be met in order to develop an effective program for controlling further drug use-related HIV infection. They include:

— The discussion of the relative success of extensive and voluntary testing of parenteral drug users for HIV infection, with consideration of whether such testing should be a precondition for receiving publicly-funded treatment services.
— The need for additional AIDS education for professionals and non-professionals involved in treating parenteral drug users.
— The need for outreach, not only to parenteral drug users but also to their spouses, sexual partners, and children.
— The need for increased funding for research, HIV testing, edu-

cation and prevention campaigns, and direct services for HIV-
infected persons.
— The need for improving the dissemination of new information
to high-risk populations and the population at large.
— The need for new drug treatment availability.

The response of state legislatures to the need for increased funds
for AIDS research and prevention has been mixed at best. According to the Intergovernmental Health Policy Project,[33] expenditures
for AIDS at the state level have grown markedly in the last few
years—from $9.6 million in FY 1984-85 to $65 million in FY
1986-87. The latter total is for the 21 states and the District of
Columbia that appropriated AIDS funds. Five states (California,
New York, Florida, New Jersey, and Massachusetts), account for
85% of the total expenditures since July 1, 1983 ($117.3 million).
Moreover, California and New York alone, the states currently
most affected by AIDS, account for 66%. Of this $117.3 million,
$5.2 million was not new money, but came from redirection or
reallocation of existing resources—usually from communicable or
sexually-transmitted disease programs.

A critical mistake will be made if future funding follows this
trend and is allocated solely on the basis of a state's current number
of reported AIDS cases. Because of the lag time between HIV infection and the development of clinical AIDS, such an approach
will inevitably lead to a rapid increase in the spread of infection.
Thus, efforts must be started or expanded immediately even in
states or cities where there are very few cases of AIDS—or perhaps
even none.

Currently, California leads the nation in providing funds for
AIDS prevention. The 1985-86 annual state expenditure for this
effort averaged 65 cents per capita; in San Francisco, however,
such expenditures are nearly eight times as high, approximately $5
per capita.[9]

We must hope that other states will follow California's lead. A
major recommendation of the Institute of Medicine[9] is that national
per capita expenditures for AIDS prevention approximate those of
San Francisco. This would require a total of more than $1 billion
annually. The Institute also suggests that the bulk of this money

come from federal sources because only the federal government has resources commensurate with the potential size of the problem.

A study conducted by the U.S. Conference of Mayors found that one of the greatest needs at the local level is funding for training and technical assistance in community education.[34] Thus far, allocation of federal funds to AIDS-related programs within PHS has been directed mostly towards basic biomedical research, vaccine development, clinical trials, and epidemiological surveillance.

A large deficit in funding of public health education and prevention efforts remains, especially for parenteral drug users. For fiscal years 1984 and 1985, less than 4% of all PHS funds were appropriated for information dissemination/public affairs. In 1986, the amount budgeted by the federal government for public education on AIDS totaled less than $25 million.[9] Any further discussion of intervention strategies for the drug-using community must be predicated on the assumption that this funding will be increased substantially.

CALIFORNIA EFFORTS

To illustrate some of the points previously made, reference to the experience in California may be useful. The state is fortunate to be early enough in the AIDS epidemic among parenteral drug users to expect significant success in combating AIDS for this group. We hope to do so by anticipating needs and allocating the necessary resources.

California's Department of Alcohol and Drug Programs (ADP) serves as the "single state agency" for alcohol and drug problems. The department funds alcohol and drug-abuse programs and coordinates alcohol and drug services.

The Executive Branch of the California government includes a Health and Welfare Agency, which represents a broad spectrum of the state's human resource programs. The state Department of Health Services (DHS) and ADP reside within and report to this agency. The agency secretary holds a cabinet position and reports directly to the Governor.

The state's infrastructure works very efficiently in terms of policy making and can implement responses to identified problems quickly once an executive decision has been made.

Several state-level AIDS advisory groups exist, providing advice to the ADP Director, the DHS Director, and the Governor. These groups have been instrumental in serving as a conduit for conveying the most recent findings from CDC, PHS, and other public and private entities involved with AIDS research, policy, or services.

ADP has also been deeply involved in policy considerations at the federal level. Governor Deukmejian has shown great interest in supporting AIDS efforts. Consistent with that interest, I have, as director of ADP, expanded my involvement in national AIDS efforts. Presently, I serve as First Vice President of the National Association of Alcohol and Drug Abuse Directors (NASADAD), and as co-chair of the AIDS and IV Drug Use subcommittee. From these positions, I have participated on NIDA's AIDS Advisory Panel and in NIAAA's forum on the Effects of Alcohol on the Immune System. Moreover, ADP has a full-time staff member responsible for AIDS efforts and subsidizes the production and distribution costs of the newsletter of the International Working Group on AIDS and IV Drug Use through its Drug Abuse Information and Monitoring Project, located at UCLA.

FUTURE CALIFORNIA AIDS EFFORT

The centralization and coordination of AIDS efforts that were emphasized earlier is occurring in California via an annual state AIDS plan. The second annual plan, for 1987-88, is now in development and serves as a vehicle for increasing the liaison among relevant state agencies, prioritizing needs, and providing contingency plans for anticipated developments. In addition, policies recommended by the plan are available for implementation in less time than has been the case in the past.

An important aspect of the plan is the consideration of how state agencies will respond to the AIDS problem after researchers have developed vaccines that will prevent AIDS or therapies that will cure it. Some of the problems we have discussed will not change: for example, it will almost certainly remain necessary to continue educational efforts aimed at preventing high-risk behaviors. Other needs will change quite dramatically following the discovery of effective treatment. For example, the problems involved in the manufacture, distribution, the appropriate prescription of and payment for medications are quite different from the problems involved in education and prevention. Payment for effective treatment will raise considerable controversy as evidenced by that surrounding the recent FDA approval of ·AZT (Retrovir), which currently costs $8,000 to $10,000 per person per year.

The development of effective therapies will make identification of infected cases at the earliest possible point paramount in importance, particularly in groups that may be removed from society's mainstream: IV drug users, prostitutes, prisoners, and arrestees. We must plan ahead so that the infrastructure already in place can adapt to handling these problems as they arise. If we can achieve this goal, then in the not too distant future AIDS may well join smallpox as a disease of the past.

State agencies must be aware of the opportunities that have emerged from the federal government's Anti-Drug Abuse Act of 1986. While funds were authorized in October, 1986, they did not become available at a state level until the summer of 1987. Agency planning must ensure that these funds and those from other sources are used to improve and extend the infrastructure already in place to combat AIDS in parenteral drug users. Future funding priorities must also be considered now, and mechanisms of obtaining and delivering them in a timely manner established. California has always been a trendsetter and I know that we will continue this tradition and help lead others in finding the most effective ways to meet the challenges posed by AIDS.

NOTES

1. Ginzburg HM, French J, Jackson J, Hartsock PI, MacDonald MG, Weiss SH. Health education and knowledge assessment of HTLV-III diseases among intravenous drug users. Health Education Quarterly. 1986; 13:372-82.

2. Morgan WM, Curran JW. Acquired Immunodeficiency syndrome: Current and future trends. Public Health Reports. 1986; 101:459-65.

3. Surgeon General. Surgeon General's report on acquired immune deficiency syndrome. JAMA. 1986; 256:2784-9.

4. Anglin MD, Brecht ML. Aids virus exposure behavior in long-term addicts: A preliminary study. Problems of Drug Dependence 1986. Proceedings of the 48th Annual Scientific Meeting, 1987, in press.

5. Ginzburg HM. Intravenous drug users and the acquired immune deficiency syndrome. Public Health Reports. 1984; 99:206-12.

6. Marmor M, DesJarlais D, Friedman S, Lyden M, El-Sadr W. The epidemic of acquired immunodeficiency syndrome (AIDS) and suggestions for its control in drug abusers. Journal of Substance Abuse Treatment. 1984; 1:237-47.

7. Curran J. Personal communication, February 9, 1987.

8. Public Health Service Plan for the Prevention and Control of AIDS and the AIDS Virus. Report of the Coolfont Planning Conference, June 4-6, 1986.

9. Confronting AIDS, Directions for Public Health, Health Care, and Research. Washington, D.C.: National Academy Press, 1986.

10. Green J, Singer M, Wintfeld A. The AIDS Epidemic: A projection of its impact on hospitals, 1986-1991. New York: University Medical Center.

11. American Hospital Association. Hospital Statistics, 1984.

12. Stall R, McKusick L, Wiley J, Coates TJ, Ostrow DG. Alcohol and drug use during sexual activity and compliance with safe sex guidelines for AIDS: The AIDS behavioral research project. Health Education Quarterly. 1986; 13:359-71.

13. Intergovernmental Health Policy Project. A summary of AIDS laws from the 1986 legislative sessions. Washington, D.C.: The George Washington University, 1986.

14. Altman D. AIDS in the Mind of America. New York: Anchor Press/Doubleday, 1986.

15. West J. personal communication, February 6, 1987.

16. Boles J. personal communication, February 6, 1987.

17. DesJarlais D, Tross S, Friedman S. Behavioral changes in response to AIDS. In: Wormiser GP, Bottone E, Stahl R, Eds. AIDS and other manifestations of human immunodeficiency infections. Parkridge, NJ: Noyes Publications, in press.

18. Moss AR. AIDS in IV drug users. Paper presented at the New Jersey State Health Department Meeting on AIDS in the IV Drug Using Community, 1986.

19. Little J. personal communication, February 10, 1987.

20. Anglin MD, Hser Y, Booth MW. Sex differences in addict careers: Treatment. American Journal of Drug and Alcohol Abuse. 1987; 13:in press.

21. National Institute on Drug Abuse (NIDA) AIDS Priorities. Technical Review Meetings on Civil Commitment for Drug Abuse, January 26-27, 1987, Rockville, M.D., 1987.

22. Covell R. World Health Organization Meeting, Stockholm, 7-9 October 1986, Summary Report. International Working Group on AIDS and IV Drug Use Newsletter. 1986; p. 1.

23. Cohen S. Some speculation about AIDS and drugs. Drug Abuse and Alcoholism Newsletter. 1985; p. 1-3.

24. Faltz B. personal communication, February 18, 1987.

25. Anglin MD, Brecht ML, Woodward SA, Bonett DG. An empirical study of maturing out: Conditional factors. International Journal of the Addictions. 1986; 21:233-46.

26. Friedman SR, DesJarlais DG, Sotheran JL, Garger J, Cohen H, Smith D.

AIDS and self-organizations among intravenous drug users. International Journal of the Addictions, in press.

27. DesJarlais DC, Friedman SR. An overview of AIDS among intravenous drug users: epidemiology, natural history and prevention. Proceedings of ICAA Drug Institute. Amsterdam: Nordwyker Houd, in press.

28. Applebaum JN, (Ed.). The AIDS Record. 1987; 1:12.

29. Wiebel W. Recommendations for the Illinois AIDS Interdisciplinary Advisory Council on Issues Relating to Substance Abuser Risk Groups. Paper presented for the Governor's Task Force on AIDS, Chicago, IL, 1986.

30. Thornton TG, (Ed.). AIDS Alert. 1986; 6:101-16.

31. Newmeyer JA. Drug abuse in the Bay Area. Unpublished Manuscript, 1985.

32. Smith DE, Chappel JN, Griffen JB, Russell KS. The practicing physicians' view of the alcohol and drug abuse patient: A training and consultation model for analyses and change. In: Smith DE, Anderson SM, Buxton M, Gottlieb N, Harvey W, Chung T, (Eds.). A multicultural view of drug abuse. Proceedings of the National Drug Abuse Conference. Cambridge: MA; Schenckman, 1977: 246-262.

33. Matthews WC, Booth MW, Turner JD, Kessler L. Physician's attitudes toward homosexuality — Survey of a California county medical society. The Western Journal of Medicine, 1986; 144:106-10.

34. Lee P, Arno P. The federal response to the AIDS epidemic. Health Policy. 1986; 6:259-67.

Treatment of Substance Abuse in Patients with HIV Infection

Barbara G. Faltz, RNC, BSN
Scott Madover, MFCC

SUMMARY. The dual diagnosis of AIDS and substance abuse raises serious clinical and ethical issues for health care providers. Often, there are barriers to the diagnosis and referral for substance abuse treatment in people with HIV infection. Countertransference is one such barrier. Important educational needs of patients can be overlooked or not fully addressed. Essential information needs to be conveyed, regardless of whether or not a patient seeks substance abuse treatment. Early intervention and treatment are essential to minimize risk for HIV infection and transmission to others. Specific clinical issues that practitioners often address for patients with AIDS or ARC are appropriate interventions for denial of the HIV-related diagnosis coupled with the denial of substance abuse, difficulties in pain management, the difficulties of family and loved ones, the need for substance abuse relapse prevention, and the need for coordination of care among agencies.

Current epidemiological data confirm the connections between substance abuse and AIDS. Seventeen percent of AIDS cases are attributed to transmission by intravenous drug use and an additional 8% involve homosexual and bisexual men who also have a history of intravenous drug use.[1] As the epidemic unfolds, the proportion of AIDS cases involving IV drug use is expected to grow dramatically. In addition, care providers are becoming increasingly aware of the multiple connections between AIDS risk and substance abuse.[2]

Barbara G. Faltz, RNC, BSN, and Scott Madover, MFCC, are with the University of California, San Francisco AIDS Professional Education Project.

These include sexual transmission of HIV to partners of intravenous drug users, neonatal transmission by infected mothers who are IV drug users or partners of IV drug users, increased risk due to disinhibition under the influence of drugs or alcohol, increased risk due to immunosuppression caused by drug or alcohol use,[3,4] and inability to utilize resources (social, financial, or health) because of substance abuse.

This paper discusses issues that arise in the treatment of people infected with HIV who also have substance abuse problems. The topics include: barriers to recognizing substance abuse in HIV-infected patients, the need for education and prevention, the need for assessment of substance abuse and HIV risk and an outline of issues involved in combining substance abuse treatment with AIDS prevention and treatment. Case vignettes drawn from the authors' experience in the AIDS Substance Abuse Program (ASAP), at the University of California, San Francisco AIDS Health Project located at San Francisco General Hospital illustrate these points.

BARRIERS TO TREATMENT

Substance abuse in a person with HIV infection or a diagnosis of AIDS or ARC can have profound consequences. The patient and loved ones are often traumatized by the diagnosis and loved ones may be reluctant to confront the patient's substance abuse. Physicians and other health workers also may overlook substance abuse in persons diagnosed as having AIDS or ARC. They may perceive the disease to be their most immediate concern and the substance abuse an ancillary issue for which a referral may be offered.

The combined problem of AIDS and substance abuse raises critical clinical, ethical, and personal dilemmas for patients, loved ones, and health care providers. The perplexities of this situation often lead to barriers to effective substance abuse treatment for individuals at risk for HIV infection, and/or diagnosed with AIDS or ARC.

The "Why Bother?" Mentality

Because of the overwhelming nature of an HIV related diagnosis, treatment providers as well as patients often feel a sense of hopelessness that precludes the possibility of major positive life changes. Rationales for ignoring drug or alcohol abuse and the possibility of successful treatment for it include: that the patient will die anyway, that substance abuse treatment will take away a "coping mechanism," and that substance abuse treatment is stressful. These rationales imply the question "Why bother treating substance abuse in patients with AIDS?"

This reasoning implies either that the quality of life is a moot issue once an individual has an HIV related diagnosis, or that somehow substance abuse will make it easier for the patient to endure his or her illness. What treatment providers may not realize is that the crisis brought about by this diagnosis may provide an opportunity to make a powerful intervention in an individual's substance abuse and that the patient's quality of life may improve as a result.

For an individual troubled by the chaotic lifestyle that often accompanies drug or alcohol abuse, the crisis evoked by an AIDS diagnosis may generate a willingness to ask, "What will I do with the rest of my life?" The patient may choose either to get treatment for the abuse or to continue it. Many do choose to "die high." However, health care providers can play a positive role in facilitating conscious choices and in making avenues of treatment available to the patient. Too often, however, those providing care for patients with HIV related risk or infection are not working in coordination with substance abuse providers. As a result, they sometimes overlook the importance of both kinds of treatment for the patient's well-being. The following case is an example of such shortsightedness:

> John L is a 38-year-old homosexual man who drank up to a fifth of alcohol daily for three years prior to admission to the hospital for pneumocystis pneumonia. He had numerous financial, relationship and vocational problems as a result of his alcohol use. He decided to go to an alcoholism treatment center after he was able to make the connection between these

problems and his drinking, and to admit to himself that he had
lost control over drinking. When his medical team heard of his
decision to enter alcohol treatment, they tried to discourage it.
They thought it would "add too much stress" to his life. However, John did go for treatment and has remained alcohol and
drug free for two years. He feels that he has "finally begun
living" and refers to his active drinking days as his "life in the
shadows."

Countertransference

Another barrier to substance abuse treatment among HIV infected patients is the countertransference of treatment providers'
fears and perceptions onto their patients. The following have been
identified as countertransference issues in treatment providers of
AIDS patients: fear of the unknown, fear of contagion, fear of death
and dying, fear of homosexuality, denial of helplessness, over-
identification with the patient, anger, and the need for professional
omnipotence.[5,6] These difficulties with countertransference issues
are magnified when providers face AIDS patients with substance
abuse problems. Issues may include fear of people with addictions,
fear of confrontation, fear of rejection by the patient, fear of loss of
control over the treatment process, or of their own or a loved one's
drug or alcohol use.

These difficulties are surmountable if treatment providers become aware of their attitudes and feelings about addiction in general
and towards individual patients who are abusing drugs or alcohol.
This self-awareness can help providers transcend their own psychological issues in providing care to HIV affected patients with substance abuse problems.

Additional Barriers

In addition to the "why bother" mentality and countertransference, additional barriers inhibit HIV affected patients with substance abuse problems from receiving adequate treatment. One is a
moralistic attitude towards alcohol and drug abuse. This attitude is
based on the unreasonable assumption that an individual is in control of their excessive use of these substances. The notion that the

addict would stop if they really wanted to can easily be used as a rationale for not aggressively encouraging treatment for the patient. In addition, drug or alcohol abuse may be considered a secondary problem of an underlying emotional illness rather than a primary condition that can cause emotional problems. This view may lead a care provider to suggest a mental health system referral rather than a more appropriate substance abuse treatment referral. Substance abuse may also be seen as the result of socio-cultural, economic or current environmental factors. This may result in referral to a social service agency rather than substance abuse treatment. A social worker may help bring about change in the patient's environment, but this change alone will not result in a major change in the addiction process.

Another ideology about recovery from addiction that can get in the way of proper treatment is the notion that a person must "hit bottom" before an effective intervention can take place. This can result in treatment providers assuming that if the patient can somehow hold their daily life together, the substance abuse problem "hasn't gotten bad enough yet."

Finally, there is the fear that the provider-patient relationship would be damaged by pointedly dealing with the substance abuse problems in HIV affected patients. The exploration, assessment and/or referral of a substance abuse problem may initially cause some stain on the provider-patient relationship. When handled in a caring, professional manner and delivered in a direct and non-judgmental way, such interventions are often very effective.

It is often true that a patient is secretly hoping to get help for their substance abuse and would welcome an offer of treatment. The goal in initial substance abuse assessment and counseling is to bring any potential problem "out in the open" and to give the patient options for change. It is important to remember that after the evaluation of substance abuse and referral for treatment, it is up to the patient to choose his or her course of action and to be aware of the consequences. In many situations a patient will continue to drink and use drugs because overcoming substance abuse is indeed difficult. One drug abuse counselor stated "The drugs always vote and often win." This can be frustrating to a health care provider who has the best interests of the patient at heart and who wonders why a substance abuser does not accept treatment. Helping professionals, as well as laypersons, require reminders about the dynamics of addiction and the need for self-motivation in recovery.

Thus far, this discussion has focused on barriers to substance abuse treatment among patients whether with a risk for AIDS or a diagnosis of HIV infection. We now turn to the need for AIDS education, prevention and treatment in substance abuse and HIV infected patients. Removing the AIDS-related barriers to substance abuse treatment, and increasing the attention paid to AIDS risk behaviors in persons with substance abuse problems, will improve the coordination and quality of care for these patients.

THE NEED FOR EDUCATION, PREVENTION AND TREATMENT

Information about the prevention of HIV infection is an essential component of all substance abuse interventions. Education should include basic AIDS information, methods of transmission, health maintenance and prevention measures.

The following is an educational guide developed in our work in the AIDS Substance Abuse Program (A.S.A.P.).[7] The "A.S.A.P. Guide to AIDS Prevention," (see Table 1), can be used at intake and in individual and group counseling sessions to educate and to problem-solve patients' risks for contracting and transmitting HIV.

EARLY INTERVENTION AND TREATMENT OF SUBSTANCE ABUSE AND HIV INFECTION

Assessment of both risk for HIV infection and for the presence of substance abuse problems need to be addressed early in treatment — regardless of which problem is the "chief complaint" in the first contact with a health care provider. Ideally, intake to substance abuse treatment includes an assessment of HIV risk behavior, a drug and alcohol history and a physical examination. Similarly, the presence of a substance abuse problem must be ruled out in clinical settings evaluating an HIV related diagnosis.

In making referrals, many options can be made available to pa-

TABLE 1 GUIDE TO AIDS PREVENTION

AIDS SUBSTANCE ABUSE PROGRAM (ASAP)

NEEDLE SAFETY:
* Don't shoot-up drugs.
* Don't share needles
* Flush works twice with household bleach (Clorox) and twice with water
* Remember that people can LOOK healthy and still carry the AIDS virus

PREVENT SEXUAL TRANSMISSION:
* Use Condoms
* Remember that people can LOOK healthy and still carry the AIDS virus.

USE OF DRUGS:
* Remember that alcohol, pot, crystal, cocaine, and poppers can lower your resistance to disease

IF YOU WANT TO HAVE CHILDREN, GET MEDICAL ADVICE IF YOU:
* Shared needles
* May have been exposed to AIDS through sexual contact.

KEEP HEALTHY:
* Get rest, exercise and a healthy diet
* Reduce stress

WANT MORE INFORMATION?
* AIDS Hotline: (415) 863-AIDS
* Drug Hotline: (415) 752-3400

tients willing to use a self-help model or a treatment program for their substance abuse problems. Health care providers who have familiarized themselves with these modalities and resources will best serve their patients' needs for information and referrals.

The HIV Antibody Test

The HIV antibody test has been proposed as part of an initial screening for HIV infection in substance abuse treatment programs. As such, it is not uniformly appropriate. If test results are received prematurely, the patient's vulnerability to relapse is enhanced. Regardless of a practitioner's skills in counseling, patients who receive positive HIV antibody test results often believe that they "have AIDS," distort or in some other way ruminate over the implications of the result.[8] Due to the anxiety that news of the results may provoke, patients frequently do not absorb the information given with test results. The providers need to be mindful of this and repeat the needed AIDS information frequently.[9,10] In addition, patients need counseling about the numerous political and social implications of revealing antibody test results.

A positive initial period of substance abuse treatment may be the most prudent first step, followed by counseling on the significance of possible test results, and the facilitation of the patient's decision regarding whether to take the test. Once revealed to an individual, an HIV Antibody test result (whether positive or negative) can provide a powerful intervention in an individual's substance abuse treatment. For some, the fear of AIDS is a "bottom" that encourages a commitment to treatment. Counselors can utilize test results by counseling patients regarding their fear of AIDS and helping the patient make connections between AIDS risk and continued drug or alcohol abuse. Counseling can also address other HIV risk behavior.

SPECIAL ISSUES IN THE TREATMENT OF SUBSTANCE ABUSE AND AIDS OR ARC

The diagnoses of AIDS Related Condition (ARC) or AIDS present dramatic emotional and social challenges for patients. These require special attention of health care and substance abuse treatment professionals. Issues of particular concern discussed below include: Denial as a stage in the grieving process (and as a continuing defense against the reality of substance abuse), difficulties in pain management, family and partner relationship difficulties, the need for relapse prevention, and the need for coordination of care.

Denial as a Stage in the Grieving Process

It is reasonable for a person to grieve about a diagnosis or risk of a life-threatening illness. As a psychological defense they may also deny that the threat is actually present. This type of denial usually continues until a person can begin to absorb the impact of the diagnosis. In the case of substance abuse, denial is a defense against seeing the extent and consequences of the addiction. Therefore, a substance abuse evaluation and intervention is most effective when a client has begun to accept their AIDS related diagnosis. In working with the patient's denial, the approach to the drug or alcohol abuse should be direct and nonjudgmental. A review of the drug and alcohol use history, together with any medical, legal, social, vocational or other resulting problems is helpful. Questioning the patient is also helpful regarding specific symptoms associated with abuse such as preoccupation with use, withdrawal, tolerance, blackouts, and use alone. This evaluation can be used to assist the patient in seeing the differences between his or her use of alcohol and drugs and social or recreational use. Denial of the substance abuse and its consequences can be very entrenched, as the following case vignette illustrates:

> Mary is a 22-year-old IV Heroin addict who was not diagnosed with AIDS when she was first seen by the authors. A child of hers had died of AIDS and she was presumed to be a carrier of HIV. When confronted with the consequences of her addiction, which were numerous and included her present hospitalization for cellulitis, Mary denied that "things were really that bad." She still shared IV needles "only with my boyfriend" and continued to have unprotected sex with him and others. She claimed that her boyfriend was in no danger of contracting AIDS because "He was strong and not gay." It was not until after her own diagnosis of AIDS that she was willing to seek substance abuse treatment.

It is difficult to predict the outcome of a substance abuse intervention with a patient. The health care or substance abuse treatment provider can find it helpful to remember that the patient education, the provider attitude, and other aspects of the interaction with the patient is remembered. Although the commitment to recovery is often not directly demonstrated, many patients, such as Mary, remember what was said and act on it months or even years later.

Difficulties in Pain Management

Successful pain management for patients with AIDS and ARC is crucial, especially when the patient has a history of substance abuse. In such patients, average doses of opiates or other psychoactive medication may not be successful in relieving pain. In addition, it is beneficial to explore alternative ways of relieving pain, insomnia, depression, and anxiety other than reliance upon psychoactive medication as the primary plan of treatment.[11] Patients who are recovering addicts or alcoholics may particularly wish to explore alternatives with their health care team. In some situations, a confrontation with a patient over the abuse of prescription medication may be called for, as in the following case:

> Joe E. was a 24-year-old man diagnosed with AIDS. At the time of contact, he had no opportunistic infection, did not have a tumor or other lesions, and was in stable condition. He appeared regularly at a medical clinic with vague complaints, lost prescriptions, bargaining for "stronger pain relief" or more tablets in each renewal. At times he was drowsy and his speech was slurred. Once he presented in opiate withdrawal. The physician confronted Joe about this situation, expressing concern that he was obtaining medications from various sources and that continued abuse of his medication would interfere with his medical treatment. This intervention led to referral for counseling and seemed to curtail further medication abuse.

It is sometimes helpful to make formal agreements with patients regarding their treatment and their use of medications. Such agreements can improve communication between the health practitioner and the patient. They can also reassure the patient that he or she will receive adequate medication for disturbing and painful symptoms.

The Difficulties of Family and Loved Ones

AIDS is a sexually transmitted disease that has primarily affected homosexual and bisexual men and IV drug abusers. All of these groups are stigmatized. Often people in these groups hide their activity and behavior from family, loved ones, employers, and other associates. This can no longer be done very successfully after an AIDS diagnosis. Often a patient is in the position of "coming out" as a gay or bisexual man or as a IV drug abuser, or both, at the time he or she reveals the AIDS diagnosis.[12] Patients may need help to be able to be as revealing as is comfortable for them about their personal life. Their desires in this matter should be respected, regardless of what they are. Health care and substance abuse treatment providers can assist in the process of problem-solving, role playing and in other supportive functions.

Family members and loved ones often share fears similar to those of health care providers discussed above. Lovers or spouses also often express "survivor's guilt" at not being the one in the couple diagnosed with AIDS or ARC. They can become hypervigilant for signs of illness, overprotective of the AIDS patient, or express a need to control environmental factors in their lives. These behaviors suggest that a partner may attempt to compensate for the overwhelming nature of their loved one's AIDS diagnosis by devoting themselves totally to the ill person. In addition, there may be the added dimension of substance abuse or co-dependency superimposed on this situation. The following is an illustration of co-dependency in a patient with AIDS who further jeopardized his health rather than confront his partner's alcoholism.

> Phil was a 36-year-old man with blindness secondary to CMV Retinitis who's partner of eight years named Alan was an alcoholic. When interviewed, he presented with a great deal of concern about his partner's drinking and the problems that it had caused. During the time they had been living together, the subject of Alan's drinking had never been addressed directly. In addition, it emerged that Phil was afraid of Alan's temper

and threats of violence when the latter was intoxicated. Alan had promised to be the primary caregiver at home to Phil. However, it appeared that Phil could not rely on meals on time or assistance from Alan with reading and other daily activities. Rather than "hurt Alan's feelings," Phil decided to go without needed assistance, thus jeopardizing his own health.

If the patient with AIDS is the substance abusing partner, the other partner may feel fear and ambivalence about confronting the substance abuse directly. They may have difficulty in being assertive about their own desires and needs. Health care providers need to be sensitive to a patient's and family member's difficulties and be aware of the increased need to control daily events and an environment now marked by trauma and uncertainty. Pre-existing co-dependency or substance abuse problems, tend to exacerbate family member's or partner's attempts to control the patient's actions and the efforts of the health care provider.

The Need for Relapse Prevention

Many difficulties face a recovering alcoholic or addict with ARC or AIDS. Alert substance abuse treatment providers can help patients plan for their response to psychological components of this diagnosis. Periods of anxiety and depression can accompany the progression of AIDS.[13] Suicidal ideation is common and needs to be assessed regularly.[14] In many treatment modalities, formal recovery plans or treatment plans can focus in on maintaining emotional health, dealing in positive ways with guilt from the past, and the spiritual and quality of life aspects of substance abuse recovery.

Two extremes in dealing with the dual diagnosis of AIDS and substance abuse are particularly dangerous. The first is the exclusive focus on substance abuse treatment issues to the exclusion of AIDS related psychological needs. A patient can be overwhelmed with grief and loss and family strife may intensify. Emotions of hopelessness and helplessness can emerge and threaten the success of substance abuse treatment. A flexible approach to treatment that allows time to discuss feelings and problems that arise from an AIDS or ARC diagnosis can serve as a vehicle for effective substance abuse treatment and for acceptance and successful coping with the life-threatening illness.

Drug use is sometimes rationalized as a way to cope with an HIV-related diagnosis. It is important that treatment providers not minimize the need for substance abuse treatment. Providers can help by continuing to confront drug or alcohol abuse. Candor and humor help in delivering this intervention in a caring and direct way. AIDS related concerns as well as substance abuse issues need to be addressed when planning for relapse prevention.

Need for Coordination of Care

Another area of difficulty in planning effective treatment for the dual diagnosis of AIDS and substance abuse is the coordination of care. Many different health care professionals, social service professionals, and substance abuse professionals may interact with one patient. Each provider has a different focus of concern and can inadvertently undermine the work of another. It is important to communicate with other members of a patient's treatment team. Drug abuse counselors can be very helpful to the health care provider in planning complementary treatment approaches, tailored to each patient's needs. An example of a successful partnership is the following case:

> Arthur E, a 32-year-old bisexual male with a history of alcohol abuse and AIDS diagnosis, was admitted to a residential alcohol and drug treatment program from the hospital. He reluctantly went to the program at the encouragement of a social worker who was concerned about his living on the streets before his hospital admission. The client got drunk on a pass from the program and was discharged because of violation of the abstinence rule. The patient enlisted several hospital departments, a social service agency and a church group to "go to bat for him." After obtaining the proper release of information, an interagency case conference was called. At this conference, a unified treatment approach was developed and subsequently followed. This coordination prevented further inter-agency conflicts and misunderstandings.

The coordination of care extends beyond the prevention of manipulation of an agency or staff member. There is the need to coordinate the use of psychoactive drugs, medical treatment, mental health referrals and social service referrals.

SUMMARY

Some of the barriers to effective dual treatment of substance abuse and AIDS-related diagnoses have been detailed and practical strategies to improve the response of treatment providers in caring for this difficult population of patients have been offered.

While there are many problems associated with the dual diagnoses of substance abuse and AIDS, health care providers and substance abuse treatment professionals can intervene effectively by first recognizing how their own values and anxieties impact upon treatment approaches. Next, they can accept substance abuse as a problem that can be treated. Finally they can stay informed of community resources to help them make accurate assessments and to develop effective treatment strategies.

NOTES

1. Centers for Infectious Diseases, Centers for Disease Control. Acquired immunodeficiency syndrome (AIDS) weekly surveillance report-United States AIDS activity. 1987:1/26.

2. Faltz BG, Madover S, Substance abuse as a cofactor for AIDS. In: McKusick L. ed. What to do about AIDS: physicians and health care providers discuss the issues. Los Angeles: University of California Press, 1986:155-162.

3. Newell GR et al, Risk factor analysis among men referred for possible acquired immune deficiency syndrome, Preven med 1985:14:81-91.

4. Mc Gregor RR, Alcohol and immune defense. JAMA 1986: 256:1474-9.

5. Dunkel J, Hatfield, S. Countertransference issues in working with persons with AIDS. Social Work. 1986:114-117.

6. DG Ostrow, TC Gayle, Psychosocial and ethical issues of AIDS health care program. Quart Rev Bltn, J Quality Assur 1986:8:284-294.

7. BG Faltz, Madover S. op. cit.

8. M Gold, N Seymour, J Sahl. Counseling HIV Seropositives. In: McKusick L. ed. What to do about AIDS: physicians and health care professionals discuss the issues. Los Angeles: University of California Press, 1986:103-110.

9. RL Binder, AIDS antibody tests on inpatient psychiatric units, AM J Psych: 1987:144:2:176-181.

10. National Institute of Mental Health, Coping with AIDS. Rockville, MD: U.S. Department of Health and Human Services. 1986.

11. RV Brody. Pain management in terminal disease. Focus. 1986:1:1-2.

12. JS Mandel, Psychosocial challenges of AIDS and ARC: clinical and research observations. In: McKusick L. ed. What to do about AIDS: physicians and health care professionals discuss the issues. Los Angeles: University of California Press, 1986:75-86.

13. JW Dilley, Treatment interventions and approaches to care of patients with acquired immune deficiency syndrome. In: Nichols S, Ostrow D, eds. Psychiatric implications of acquired immune deficiency syndrome. Washington: American Psychiatric Press, Inc. 1984:62-70.

14. P. Goldblum, J Moulton. AIDS-related suicide: a dilemma for health care providers. Focus. 1987:2:1-2.

15. JR Acevedo, Impact of risk reduction on mental health. In: McKusick L. ed. What to do about AIDS: physicians and health care professionals discuss the issues. Los Angeles: University of California Press, 1986:95-102.

Current and Future Trends
in AIDS in New York City

Stephen C. Joseph, MD, MPH

SUMMARY. Over the past five years, the proportion of AIDS
cases among gay and bisexual men in New York City has fallen as
the proportion of AIDS cases has risen among IV drug abusers, the
major channel of HIV infection to children and heterosexuals in
New York City. By 1991, there will be over 40,000 cumulative
AIDS cases in New York City, with close to 30,000 deaths. Controlling the spread of HIV infection among drug addicts, children, and
heterosexuals requires education, outreach, and exploration of strategies such as increasing the legal availability of sterile needles and
syringes. For at least the next several years, massive programs of
public health education and voluntary, confidential risk-reduction
counseling and HIV antibody testing will remain our critical weapons for reducing the spread of HIV infection among people engaged
in high-risk behavior as well as among the general public.

Nowhere in North America is the relentless tragedy of the AIDS
epidemic more starkly felt than here in New York City. Since 1981,
over 9,700 cases of people with AIDS have been diagnosed here—
30% of the national total. Over half of these people have died.
AIDS is the leading cause of death in New York City among men
aged 25-44 and women aged 25-29.

Gay men account for the largest percentage of reported cases in
New York City. (See Chart 1.) Over the past five years, the proportion of cases of gay and bisexual men has fallen from 73% to 55%,
as the proportion of cases among IV drug abusers has risen, from

Address for reprint requests: Stephen C. Joseph, MD, MPH, Commissioner of
Health, 125 Worth Street, New York, NY 10013.

159

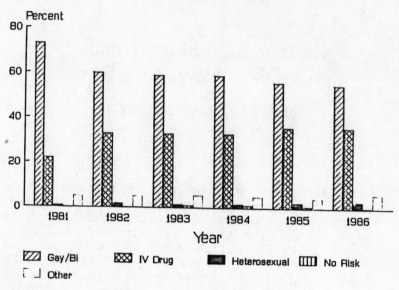

CHART 1. Percent distribution of AIDS in New York City by patient group and year of diagnosis. Source: New York City Department of Health.

22% to 36%. Up to 60 percent of New York City's estimated 200,000 heroin addicts are thought to be HIV-infected.

Thirty-one percent of AIDS cases in New York City are among blacks; 23% are among Hispanics. (See Chart 2.) Together, these groups represent over half of all people with AIDS in New York City. Eighty-six percent of male IV drug users with AIDS are black or Hispanic; 90% of mothers of children who have AIDS are black or Hispanic. The high proportion of blacks and Hispanics with AIDS reflects the link between AIDS and poverty and drug abuse in New York City.

Surveillance of gay-related sexually transmitted diseases reveals that since the onset of the epidemic, the rates of pharyngeal and rectal gonorrhea among gay men have declined, while gonorrhea rates among women and heterosexual men have increased. (See Charts 3 and 4.) This indicates that many in the gay community seem to have incorporated prevention information into their behavior to decrease the chances of transmitting the virus. Our attention

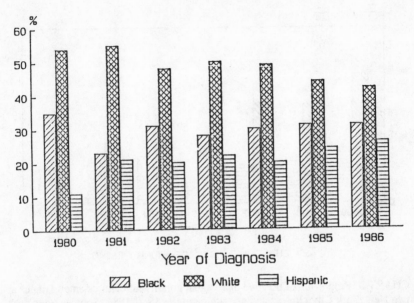

CHART 2. AIDS incidence in New York City by race/ethnicity. Source: New York City Department of Health.

turns increasingly to the IV drug abuser, the major channel of infection for babies born to infected mothers and heterosexuals in New York City.

Chart 5 shows the rising numbers of women of child-bearing age and children who have AIDS (excluding transfusion-associated cases). Of the 183 reported cases of children with AIDS in New York City, IV drug use by one or both parents is involved in over 80%. (See Chart 6.) Between 300 and 800 HIV-infected children are estimated to be born in New York City each year.

Of the 975 women with AIDS diagnosed in New York City since 1981, 80% have been IVDAs or the sex partners of IVDAs. Studies have shown conclusively that heterosexual transmission of the AIDS virus does occur through vaginal intercourse. Important questions remain, however, about the relative efficiency of various types of sexual transmission. The data so far in New York City indicate far more efficiently transmitted male-to-female infection than female to male. (See Chart 7.) Of the 227 cases of heterosex-

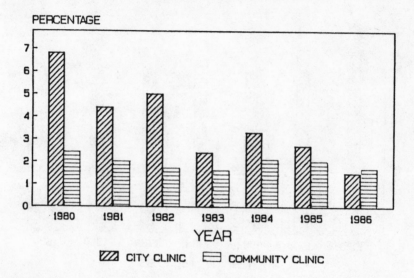

CHART 3. Proportion of positive pharyngeal gonorrheal cultures among males at two New York City clinics, 2nd and 4th quarters, 1980-1986. Source: New York City Department of Health.

ually transmitted AIDS that Health Department surveillance statistics have identified in New York City to date, 222 have been women. Because New York City has over half of all infected female cases of AIDS in the United States, if female-to-male heterosexual transmission were as efficient as male-to-female transmission, we would expect to see heterosexually transmitted HIV infection in larger numbers of men, and we have not to date.

In 1981, the first case of heterosexually transmitted AIDS was identified in New York City, representing 0.8% of the total number of 1981 cases. The number of cases of AIDS whose risk has been heterosexual contact with a person in an AIDS risk group has increased from 2% (8 cases) in 1982 to 3% (71 cases) in November of 1986. Over 90% of the heterosexually transmitted HIV infection is from sexual contact with male IV drug users, and the proportion of IV drug users with AIDS has increased from 22% in 1981 to 36% currently. If bidirectional heterosexual transmission is more efficient than we predict from available data, the source has been and

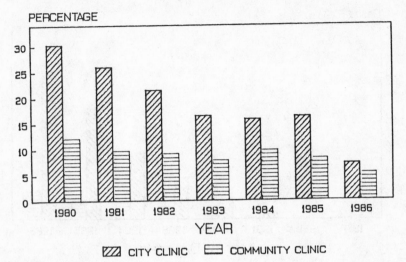

CHART 4. Proportion of positive rectal gonorrheal cultures among males at two New York City clinics, 2nd and 4th quarters, 1980-1986. Source: New York City Department of Health.

will probably continue to be from the population of infected IV-DAs.

During the last five years, we have seen a dramatic increase in the numbers of deaths of IV drug abusers. (See Chart 8.) They have been dying from TB, pneumonia, and other conditions that could very well be complications of AIDS-related immunosuppression not identified as CDC-defined AIDS. For example, in 1985, 92 addicts died of endocarditis, compared with only 31 in 1980; in 1985, 252 addicts died of nonspecific pneumonia, compared with 39 in 1980. During this time, drug-overdose deaths remained relatively constant, indicating no dramatic increase in the number of drug addicts. If we adjusted our surveillance for this increase in deaths among IV drug addicts, the absolute number of AIDS-related illness and deaths in the City would be as much as 50% higher than is currently reported.

Our own best current estimates project that by the end of 1991, there will be over 40,000 cumulative AIDS cases in New York City, with close to 30,000 deaths. (See Chart 9.) Most of these

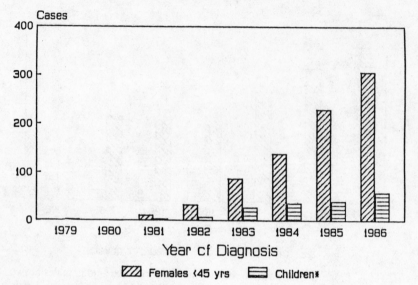

CHART 5. AIDS incidence in New York City among females below 45 years of age and children, December 1986. Source: New York City Department of Health.

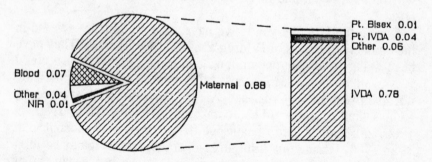

CHART 6. New York City pediatric AIDS cases by risk groups, March 1987. Source: New York City Department of Health.

people are among the estimated 500,000 HIV-infected individuals currently in New York City.

The U.S. Public Health Service projects that by 1991 the total number of AIDS cases across the country will reach 270,000, with

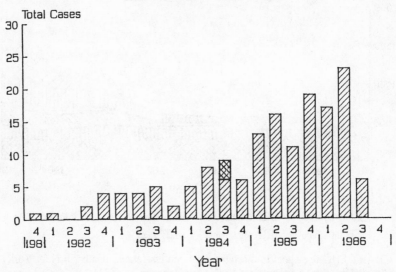

CHART 7. Heterosexual AIDS cases in New York City, 1981-1986. Source: New York City Department of Health.

179,000 deaths, as the virus spreads outside New York and San Francisco and other current areas of highest prevalence. Nationally, homosexual and bisexual men will account for more than 70% of the cases diagnosed in the next five years. Nationally, 25% of the new cases will be intravenous (IV) drug abusers. Nationally, 3,000 cases of AIDS in children will have been seen by 1991, compared with more than 400 so far. These figures include only AIDS, not other HIV-related infection.

Our projections are based on a five-year horizon. They assume no change in transmission patterns or disease expression among those presently infected. Because it is hard to know how many people are practicing high-risk behavior, the figures may be underestimates. If the proportion of heterosexually infected people, or other risk groups, increases significantly, our projections would have to be revised upwards.

No one knows the 10- or 20-year clinical outlook for those now infected. Nor do we know what long-term burdens will be posed by

Number of Deaths

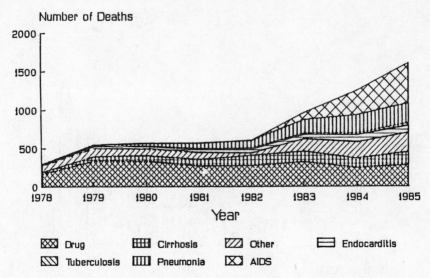

CHART 8. Cause-specific mortality in narcotics-related deaths in New York City, 1978-1985. Source: New York City Department of Health.

Cumulative Number of Cases (Thousands)

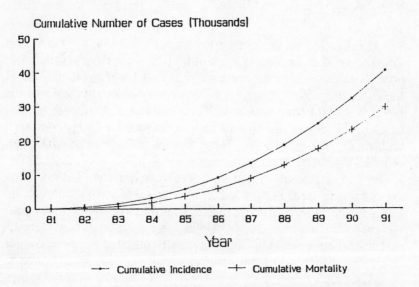

CHART 9. AIDS projections: cumulative incidence and cumulative mortality in New York City, 1981-1991. Source: New York City Department of Health.

illness associated with HIV infection, such as TB or lymphoma.
The impact of HIV-immunodeficiency on the trends of other dis-
eases is under study and as yet is incompletely understood. (See
Charts 10 and 11.)

The import of these figures is clear: By 1991, when the number
of newly diagnosed cases for that year will exceed our current cumulative cases to date, the impact of the AIDS epidemic on our
citizens, our hospitals, and the entire city will be beyond that of any
public health crisis of modern times.

In New York City, the common element in the transmission of
AIDS among addicts, children, and heterosexuals is the practice of
needle-sharing. Applying traditional public health measures to control the spread of infection in drug users is difficult. We lack the
risk reduction and community responsibility models developed by
the gay community to ensure that our health promotion and education efforts are effective.

To more adequately address the problem of needle-sharing

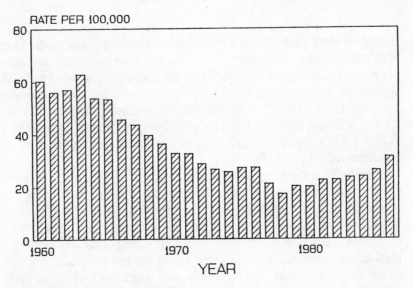

PROJECTED BASED ON 10 MONTHS

CHART 10. Tuberculosis morbidity rates in New York City, 1960-1986.
Source: New York City Department of Health.

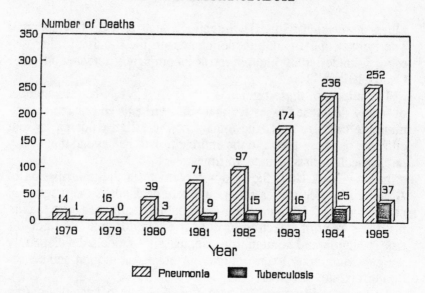

CHART 11. Pneumonia and tuberculosis mortality among narcotic users in New York City, 1978-1985. Source: New York City Department of Health.

among IV drug users, New York City has a number of options. One is recommending abstinence from self-injection of drugs, and providing increased access to treatment for addicts. To promote AIDS awareness and prevention among IV drug users and their sex partners, the Health Department distributes brochures, flyers, and wallet and subway cards in English and Spanish. Health Department Public Health Educators are assigned to local areas with highest incidence of IV drug use.

In ongoing studies at methadone treatment clinics in New York City, up to 95% of seropositive IV drug users who have undergone treatment for their addiction have stopped using needles. Yet New York City already has more demand for drug abuse treatment than treatment capacity. Currently, there are long waiting lists for methadone maintenance; drug-free rehabilitation programs are full to capacity. A person must typically wait weeks or months to enter a drug treatment program. During this time, he or she is almost certain to continue to inject drugs and possibly contribute to the spread of HIV. The federal government must consider drug rehabilitation programs to be one of our important priorities, and provide more federal dollars through the states and communities so more addicts may find treatment.

Another option is getting people to use clean drug injection equipment, or "works." The Health Department has produced a brochure, *Aids and Drugs*, which cautions that "the best protection is no injection," and also describes methods of cleaning works. We support many city-wide and local organizations, including the community group ADAPT (Association for Drug Abuse Prevention and Treatment), to reach those at risk through IV drug abuse. We are working with ADAPT on projects such as doing AIDS prevention and education programs in high risk areas, even within shooting galleries themselves. We also cooperate with the state program of outreach to addicts.

We need to explore additional strategies for reducing the spread of HIV among drug users. One strategy is to increase the legal availability of sterile needles and syringes. No research has measured the effectiveness of this approach in this country. New York law requires prescriptions for the sale of needles and syringes, and laws prohibit the possession of narcotics paraphernalia. (New York State is one of only 12 states that require prescriptions for the purchase of needles and syringes.)

The New York City Department of Health, in conjunction with the New York State Division of Substance Abuse Services, developed a proposal for a research project to determine the effect of the availability of clean needles to a small, carefully selected and monitored group of IV drug addicts. The design requires the State Health Commissioner's approval before the research project can actually begin. Once approved, the research study will be the basis for evaluating the efficacy of this approach in halting the further spread of the virus.

This program is a pilot study applicable to a limited number of participants, not a public health scale intervention to reduce the spread of HIV within the City as a whole. Education on risk-reduction behavior is a key element in the distribution program. Studies demonstrate that demand on the street for new needles has arisen as a response to educational messages regarding the health hazards of needle sharing. An educational campaign incorporated into the needle exchange program could lead to a considerable reduction in the sharing of needles, and thus an increased demand for new ones.

Many of the objections to the program center on the belief that increasing the availability of needles will give non-IV drug users the stimulus to begin IV drug use; increased availability of needles, it is argued, increases the number of IV drug users. Any public policy in this area needs to balance two imperatives: taking all feasible measures to slow HIV transmission, and avoiding promotion of IV substance abuse.

For at least the next several years, the most effective measure for significantly reducing the spread of HIV infection is education of the public—and not just those formerly considered to be in high-risk groups—on the need to halt risk-promoting behaviors—as well as the need to practice safe behaviors, such as condom use.

Surgeon General Koop has repeatedly said the "best protection against infection right now, barring abstinence, is use of a condom." The Health Department has undertaken a large-scale, $1 million campaign, this year, to promote latex condom use in all appropriate situations. This will include distribution of 3 million condoms in the next year, along with educational materials, wherever we have the chance to discuss the subject.

We have recently joined with the Correction Department in a broad-based education and risk reduction program to educate jail inmates about AIDS and provide condoms to those at highest risk of infection. One-quarter of the 100,000 inmates who pass through New York City jails each year are estimated to be HIV-infected. We must recognize that sex in prisons may help spread AIDS inside and outside the prison walls.

The Surgeon General's call for frank, explicit discussions about the consequences of sexual behavior is a long-overdue national recognition of the severity of the problem, and underscores people's need for comprehensive education and counseling. Yet, because AIDS involves two of the most personal areas of human behavior—sexuality and drug use—society in general is not comfortable discussing certain AIDS issues, such as needle exchange and condom use. Some segments of society fear that advocating sex education and safer drug and sex behavior, including condom use, will be interpreted as encouraging promiscuity and promoting homosexuality and drug use. This is foolishness of the first order. We are facing a public health crisis of immense magnitude. With information and education our only current effective tools, we must squarely face those facts we know, be honest about our areas of uncertainty, and use explicit, understandable language repeatedly and clearly. Despite those who would hope the problem would disappear if we ignored it, health professionals, educators, and the media have an obligation to confront the problem head-on. This is the best, and currently only, way to protect our society.

As the epidemic deepens, pressures are mounting for measures that many public health experts believe should not currently be taken. These include universal or mandatory HIV antibody screening programs, as well as isolation and quarantine, traditionally used to contain contagious diseases.

Our nation's leading public health officials overwhelmingly oppose mandatory AIDS blood tests. Virtually all public health officials in areas of highest prevalence agree that compelling people to learn their serostatus when no treatment is available, and confidentiality cannot be assured, would be unwise and counterproductive. Efforts to control the spread of AIDS depend upon education about risk reduction for those infected or at risk of infection; if those whom we are trying to reach with our educational messages fear social sanctions and discrimination, they will not come forward.

New York City is moving aggressively to make voluntary and confidential counseling and testing much more widely accessible. The Department of Health policy on the increased availability of voluntary, confidential counseling and HIV antibody testing contains guidelines for health professionals on preventing HIV infection, and educating those at risk.

Anyone in New York City should be able to know his or her antibody status *provided* the test results are confidential, tests are voluntarily undergone, and counseling is available before and after testing. We recommend that physicians *actively* consider if their patients may be at risk of HIV infection, discuss risk and risk-avoiding behavior as a part of routine medical care of all patients, and offer counseling and, if appropriate, HIV testing to patients at risk.

Testing in New York City is available through free, anonymous test sites established in the City by the City and State Health Departments; through any licensed physician in New York City, using a laboratory with a special permit issued by the New York City Health Department if counseling, consent, and confidentiality are guaranteed by the lab; and at our Sexually Transmitted Disease clinics in New York City.

By the beginning of this year, over 17,000 people had been counseled and tested. We will continue to make voluntary, confidential counseling and testing sites more widely available. To support the confidential, voluntary aspects of counseling and testing, the Department of Health has adopted a course of action known as contact notification. We urge, and directly and actively assist when asked, people who are seropositive to notify their sex or drug partners.

Stands on issues such as contact tracing and mandatory HIV antibody testing have at times been falsely portrayed as sacrificing the health of the public in the name of civil liberties. Yet the public health approaches taken by New York City are based on sound medical and scientific principles; virtually all public health officials in areas of highest prevalence agree that the health of the public would be hindered by control measures such as mandatory testing.

Recently in New York City there was the case of a woman with tuberculosis who was afraid to reveal her employer (which turned out to be the police department), because she was afraid of losing her job. She would not have lost her job, but health officials could not check her contacts because she withheld that information. If someone with a treatable disease is not forthcoming because she fears discrimination, HIV-infected people or those actually afflicted with AIDS or AIDS-related conditions would certainly not come forward if they feared other forms of discrimination in addition to the ones they already face. They would hide, and refuse voluntary cooperation with the education, counseling, and testing that are our main weapons against AIDS at this time.

In New York City, an interagency task force on AIDS coordinates planning and programming among different levels of government. In Fiscal 1987, New York City will spend approximately $258 million on programs in AIDS treatment, human services, public health initiatives, and human rights discrimination. The City share of this figure is $73 million in tax levy dollars. Next year, we will spend $335 million, of which $90 million is City tax levy dollars.

The National Academy of Sciences report charged that the federal response to the AIDS epidemic has been dangerously inadequate. The Academy endorsed a request for $471 million in AIDS research funds in 1988. The Academy also recommended $1 billion a year be newly appropriated for extensive basic and applied biomedical research to better understand the disease and increase the likelihood of producing a safe and effective drug or vaccine as soon as possible.

The Academy also estimated that another $1 billion would be needed annually for the massive, continuing education campaign needed to increase public awareness of the ways in which people can protect themselves against infection. The money would also be applied to other necessary public health measures, such as screening the blood supply, voluntary confidential testing, and increased efforts in treatment and prevention of IV drug use.

The Academy's call for a $2-billion-a-year education and research effort must herald an increase in the national commitment against AIDS. Until recently, the federal government left the responsibility for prevention of the spread of AIDS, as well as caring for AIDS victims, to state and local governments. The federal government must do much more.

I have touched upon only a few of the many public policy issues that face us. Among other critical issues are (a) the increasing need for coordinated planning and programming both within the City and with other levels of government to face the very large resource requirements; (b) the pressures that are mounting on our public health, human services, health care, and human rights discrimination systems; and (c) the changes in our current public health and clinical approaches that may be required as effective medical therapy becomes available.

Looming over all, of course, is the uncertainty as to whether the current rate and extent of heterosexual transmission will alter significantly.

While the recent federal approval of AZT is welcome, the drug is not a magic bullet against the disease; biomedical research will not eliminate the problem of AIDS in the short run. Education, counseling, and voluntary, confidential testing will remain our critical weapons in the fight to contain AIDS for at least the next several years. We must do all we can to advance these weapons against this major and mounting health problem in New York City, across the United States, and around the world.

The Role of
Substance Abuse Professionals
in the AIDS Epidemic

David E. Smith, MD

SUMMARY. The article examines the expanding role of drug abuse treatment professionals in the AIDs epidemic, including the utilization of predictive models to help focus prevention, intervention, and treatment efforts for IV drug users, who are the second largest at-risk group for HIV contagion after homosexual males, as well as adolescents. More specifically, the article addresses such issues as the needle-sharing ritual, the "teach and bleach" method of community-based outreach prevention, some of the cofactors involved in immunosuppression, the controversy surrounding drug maintenance programs and free needle distribution, safer-sex guidelines and sex education, HIV antibody testing and confidentiality. This article postulates that it is crucial for chemical dependency treatment programs to apprise themselves, on an ongoing basis, of both individual and organizational guidelines regarding AIDS and to develop AIDS and chemical dependency treatment components within their programs in order to keep drug abuse treatment itself from being compromised.

David E. Smith, MD, is Founder and Medical Director, Haight Ashbury Free Medical Clinics, 409 Clayton Street, San Francisco, CA 94117, Research Director, Merritt Peralta Institute, Chemical Dependency Recovery Hospital, Oakland, California and Associate Clinical Professor of Occupational Health and Toxicology, University of California, San Francisco. This paper was presented at The American Medical Society on Alcoholism and Other Drug Dependencies National Forum: AIDS and Chemical Dependency, February 21-22, 1987.

THE AIDS EPIDEMIC
AND THE INCREASING ROLE OF DRUG ABUSE
AS A RISK FACTOR

Homosexual males still represent the largest group at risk for AIDS, but intravenous drug users (IVDUs) are the next largest risk group. In the San Francisco Bay area, approximately 73% of the AIDS cases have occurred in gay and bisexual men, with 17% occurring in IVDUs. Ten percent of the gay men are also IVDUs. This percentage of AIDS cases associated with IVDUs in San Francisco is similar to the national statistics reported by the Centers for Disease Control;[1] but in New York City and New Jersey, where the highest incidence of drug-related AIDS cases is reported, IVDUs account for 34 and 54% of total cases, respectively.[2] Obviously, in an expanding epidemic these percentages will change, but the number of IVDUs involved in the AIDS epidemic is increasing, and gay men who are IVDUs appear to currently be the highest risk group.

It should be noted that certain non-IVDU practices are associated with high-risk sexual practices. For example, the Substance Abuse and Sexual Concerns Research Project of the Haight Ashbury Free Medical Clinic (HAFMC) found that phencyclidine (PCP), methaqualone (Quaalude®) and nitrites were widely used in the gay male community as part of anal-related sexual practices, including group anal intercourse and anal-manual intercourse, to obtain a level of disinhibition, reduce pain and enhance fantasy. These drugs were taken orally or inhaled, but they were not injected. Nonetheless, they did facilitate high-risk sexual practices.[3] There appears to be a high correlation between these drugs, certain sexual practices, and the development of AIDS,[4] particularly Kaposi's Sarcoma.[5]

Nationally, 25% of the total AIDS cases are IVDUs, with 8% being the crossover group of gay IVDUs. Clearly IV drug use represents an important bridge in the spread of the HIV virus to other segments of the population, including heterosexuals. There is also evidence that, as a group, IVDUs have a more virulent, rapid onset course, although this may relate to a variety of other cofactors, such as the abuse of other immunosuppressant drugs (including alcohol) and poor nutrition.[6]

PREDICTIVE MODELS

Community-based health care and drug abuse treatment programs, such as HAFMC, are getting more involved in the study of the AIDS epidemic as well as developing methods for prevention, intervention and treatment. Dr. John Newmeyer, epidemiologist and coordinator of HAFMC's AIDS Project, has developed theoretical models for predicting AIDS morbidity that have demonstrated that the AIDS morbidity curve in San Francisco is still very much on the upswing and will not start to level off for several more years (see Graph 1).

In developing this innovative epidemiological predictive method, Newmeyer emphasized that it is not just the number of high-risk invasive contacts (e.g., anal intercourse, IV drug use, blood transfusions) that contribute to the risk of getting AIDS, but it is also the number of different partners or contacts. For example, in analyzing the cumulative risk of having various sexual partners, Newmeyer stressed that safer sex leads to a lesser cumulative risk than unsafe sex, even if the former is frequent and the latter infrequent.

However, even for those men who continue to practice unsafe techniques, the risk could be analyzed as follows: Pattern A — receptive anal intercourse 140 times with one partner; Pattern B — receptive anal intercourse 70 times with each of two partners; and Pattern C — receptive anal intercourse twice with 70 partners. The frequency of contact is the same in all three patterns, but the cumulative risk with Pattern C is substantially higher than with Patterns A or B. Furthermore, this cumulative risk goes up when the reservoir of infection in the community is higher. As Newmeyer has demonstrated (unpublished data), in early 1982 (with a 6.5% infection rate), the percentages of risk for Patterns A, B and C in San Francisco were as follows: 5.2%, 8.31% and 12.85%, respectively. By early 1986 (with a 40% infection rate) the percentages for Patterns A, B and C had increased to 35.18%, 45.41% and 57.19%, respectively. Newmeyer emphasized that unsafe sex put one at greater risk for getting AIDS in 1986 than in 1982 because the background infection rate was so much higher. He also indicated that in 1982 monogamy offered a much greater reduction of risk than in

Graph 1

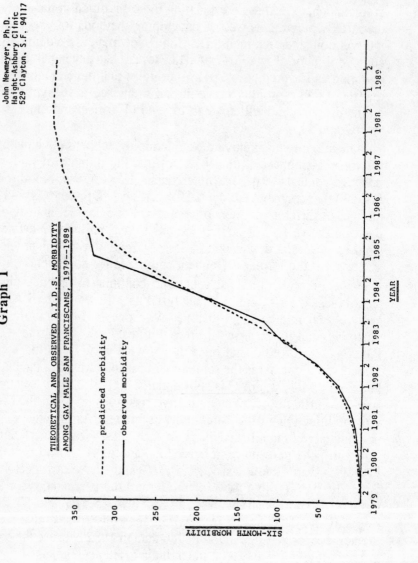

John Newmeyer, Ph.D.
Haight-Ashbury Clinic
529 Clayton, S.F. 94117

THEORETICAL AND OBSERVED A.I.D.S. MORBIDITY
AMONG GAY MALE SAN FRANCISCANS, 1979--1989

------ predicted morbidity

——— observed morbidity

SIX-MONTH MORBIDITY

350

300

250

200

150

100

50

1979 1980 1981 1982 1983 1984 1985 1986 1987 1988 1989

YEAR

178

1986, and that safer-sex practices substantially reduce the potential for infection.

Using the same model, one could also begin to develop predictive models for IVDUs in regard to the risk of acquiring AIDS: Pattern A—IVDU sharing needles 140 times with one partner; Pattern B—IVDU sharing needles 70 times with each of two partners; and Pattern C—IVDU sharing needles twice with each of 70 partners. Each pattern represents the same number of invasive IV contacts, but the frequency of needle sharing and the number of partners in Pattern C produces a substantially higher cumulative risk than Pattern A or B. And the cumulative risk of needle sharing in 1986 is substantially higher than in 1982. Obviously, the highest risk group would be those who inject drugs intravenously as part of a needle-sharing ritual that included frequent partners.

Studies of amphetamine abuse and sexual dysfunction,[7] have found that IV drug abuse and sexual marathons involving frequent anonymous contacts were part of a specific drug/sex ritual in *both* the speed and gay communities. Amphetamine reportedly enhanced sexual fantasy, prolonged erection, and interfered less with sexual function than the depressant drugs. In general, stimulant abuse seems to be on the rise, and this pattern appears to particularly be on the increase in the San Francisco Bay Area in what appears to be the highest AIDS risk group, where needle sharing is part of the sexual ritual and is an integral part of this specific drug practice. However, even if an IVDU shares needles with only one partner as part of a specific drug/sex ritual, there would be some cumulative risk of AIDS, which again would be higher in 1986 than 1982. A suggestion for future research would be to assess the cumulative risk of fellatio and masturbation, two very common gay male sexual practices—often to the exclusion of anal intercourse on a lifetime basis.

OUTREACH METHODS DESIGNED TO IMPACT ON AIDS AND IVDUs

Newmeyer and colleagues determined that the proportion of heterosexual drug abusers among all AIDS cases in the target community was approximately 17%, with an additional 9% of cases being gay men with a history of IV drug use.[8] The mode of AIDS transmission among all heterosexual IVDUs and many gay IVDUs appears to be direct blood-to-blood contact that occurs from sharing contaminated needles. Based on this theory, Newmeyer and associates have developed innovative community outreach programs to impact on IV drug abuse and the risk of AIDS, primarily targeting gay and bisexual male IV stimulant abusers as well as IV heroin users.

A community outreach and intervention strategy was developed for the prevention of HIV virus contagion among IVDUs that required an understanding of the current stage of the AIDS epidemic in the San Francisco Bay Area as well as the current state of scientific knowledge of AIDS transmission among IVDUs. In addition, the moral, legal and political climate of San Francisco had to be considered, as well as the cultural ritual of drug users and the clandestine nature of IV drug use. Newmeyer theorized that the voluntary participation of IVDUs in risk-reduction measures could be achieved despite the fact that this is a very difficult population to reach. The outreach approach followed intensive ethnographic research in San Francisco neighborhoods and discovered which ones had the largest concentrations of IVDUs. It was found that more than 90% of the interview sample had recently shared hypodermic equipment, so the potential for rapid contagion was clearly evident.

Furthermore, the studies at HAFMC involving community-based drug treatment as well as the community outreach work indicated that 16% of the IV drug using population was seropositive and was increasing in San Francisco, particularly for those not in treatment. Although this seropositive rate for IVDUs is less than has been reported in New York City, the rate of increase was particularly alarming. Newmeyer and associates determined that the large-scale shooting galleries of a commercial orientation in New York City rarely exist in San Francisco, but that ad hoc facilities in residential hotels and apartments for more informal needle sharing did exist and that needle sharing was an integral part of the IV drug use ritual. Their theory of how to interrupt IV needle-sharing AIDS contagion involved four approaches: (1) stopping drug use altogether; (2) failing to do that, cease injecting drugs; (3) if injecting drugs, not to share needles and to have one's own equipment; and (4) if sharing needles, disinfect equipment that is being shared.

Newmeyer's study indicated that there was a heightened aware-ness by IVDUs about AIDS transmission via sharing dirty needles and that a number of techniques of varying effectiveness were used to clean needles and outfits. In interviews with active IVDUs at the HAFMC (unpublished data), addicts stressed that they were much more careful about needle sharing than their predecessors, the "speed freaks and junkies" of the 1960s and 1970s. They stressed that their needle sharing was much more controlled, except when they "got too stoned or junk sick" (opiate withdrawal). It is impor-tant to mention that the San Francisco Bay Area may have different drug rituals than other parts of the country, given that agencies like the San Francisco AIDS Foundation and HAFMC are very visible and may have an enormous impact on these rituals via educational programs over the past few years. In the same fashion that studies of IVDUs and hepatitis in the latter 1960s and early 1970s indicated that the population at risk understood hepatitis transmission via nee-dles, there could be some alteration of needle-sharing rituals while still continuing to participate in the ritual.

Ultimately, the outreach plan consisted of three phases: (1) an education effort that clearly associated the spread of AIDS with the sharing of contaminated hypodermic needles; (2) direct advice and counseling on safe-needle and safer-sex practices that protect the at-risk population; and (3) aggressive monitoring of the IVDUs in or-der to document and encourage compliance with safe-needle and safer-sex practices.

THE YOUTH ENVIRONMENT STUDY PROJECT AND BLEACH BOTTLE DISTRIBUTION

In conjunction with community outreach work, the Drug Detoxi-fication and Aftercare Project of HAFMC also provides free care to IVDUs in San Francisco in an attempt to both educate them about the risk of IV drug abuse and to treat their addictive disease so that they may effectively extricate themselves from the drug- and nee-dle-using environment. From a clinical perspective, the approach combines both a prevention and a treatment strategy aimed at the highest-risk groups, with an emphasis and commitment aimed at reaching IVDUs and those people who have the fewest resources to seek help. To implement this program, HAFMC developed a brochure in 1983 titled "Shooting Up and Your Health," which was directed at IVDUs and provided information about several diseases, including hepatitis, endocarditis and AIDS.

In 1985, the educational approach shifted somewhat to "Don't Share Needles," and became the basis for a much larger campaign that included the San Francisco AIDS Foundation and HAFMC. By way of the Youth Environment Study (YES), Newmeyer and associates developed an outreach method to introduce the concept of effective cleaning procedures in the needle-sharing ritual, and an educational technique to carry that tactic to IVDUs so that its adoption would have a high probability of effectiveness. The major outreach criteria were that the materials and the method had to be effective in destroying the HIV virus, that materials would be inexpensive (because IVDUs tend to have little discretionary income), that the method had to be quick and convenient because generally the needle-sharing group has a limited supply of hypodermic needles, and that the method must be safe so that the IVDU does not expose himself or herself to additional substantial risk by its application. What evolved has come to be known as the "teach and bleach" method, which utilizes one ounce YES bottles labeled "Household Bleach" ("Chloro" in the Spanish version) and with instructions clearly printed on the bottle. This small amount of bleach is enough for 20 rinses. (See Figure 1.)

Using the teach and bleach method as well as handing out both the small bottle of beach and condoms, community outreach workers serve a number of functions. First, they deliver the safe-needle message directly into the hands of the at-risk population as well as giving IVDUs a convenient container that they can keep for refills. Second, they introduce a novel aspect of the needle-sharing ritual that helps to make it less automatic and unconscious. By having to think and talk about this new aspect of the ritual, it becomes a more conscious act. This may trigger discussion of AIDS among IVDUs and help them to take positive action to protect themselves from the AIDS virus. All this accelerates the movement of prevention information within the drug-using network and enhances the image of the outreach workers as being caring and nonjudgmental, and it establishes a rationale for continued contact with the outreach work-

prepared by the Haight Ashbury Free Medical Clinics

FIGURE 1

183

ers that permits the development of a positive and sustained relationship.[9] By having outreach workers going into the homes, hotels and entertainment hangouts of IVDUs, an alliance is formed that not only allows AIDS prevention education to be disseminated, but also facilitates instruction about safer-sex practices and condoms. Moreover, access to drug treatment facilities is significantly increased.

It should be noted that this method of combining prevention of IV drug use and AIDS contagion with community-based treatment for addictive disease has been controversial in many circles, but nevertheless it demonstrates a balanced effort that is based on certain assumptions: to interrupt the spread of the AIDS virus while encouraging a treatment model for IVDUs that emphasizes nondrug alternatives to a high-risk lifestyle.

INDIVIDUAL EDUCATION AND TREATMENT IN CHEMICAL DEPENDENCY PROGRAMS

When the IV drug abuser comes into treatment at HAFMC, the client receives management of medical complications, detoxification from the drug of dependence, and specific education about addictive disease and the associated medical complications, including the risk for AIDS. Also, there is individual counseling and exposure to the group recovery process, which increases awareness of 12-Step programs — such as Alcoholics Anonymous, Cocaine Anonymous and Narcotics Anonymous — and there is also exposure to family treatment and psychotherapy if indicated. Medication may be used during the detoxification stage, but the goal of treatment is to learn to live a comfortable and responsible life without the use of psychoactive drugs.

However, there is a prevailing notion among a significant number of health care professionals who are not versed in the chemical dependency treatment field that IV drug addiction is not a treatable entity (and that needle-sharing rituals do not exist). Clearly, if one believes this then a logical goal would be to have mass distribution of clean needles and a legal supply of drugs, rather than focus on drug-free recovery. The AIDS epidemic is having a marked impact on the selection of chemical dependency treatment modalities in the

public sector, and greater emphasis is being placed on drug maintenance rather than drug-free treatment modalities.

Forest Tennant, MD, a leading drug abuse expert, has stated that
there is "a terrible bias about narcotic substitution, and . . . a naive
idea that we can cure narcotic addicts in great numbers, and we
can't." Dr. Tennant stressed that in the next six months he would
like to see every intravenous drug addict in the country given the
chance to take methadone, LAAM, or whatever they will take to get
them off needle use. Dr. Tennant did state that

> We may be looking at a greater risk than nuclear war, because
> we may all die from this disease. Historically, there may have
> been races which have disappeared, and we don't know what
> happened. Possibly a virus knocked them out.[10]

Selwyn and associates have demonstrated that both education and
methadone maintenance can reduce needle-sharing behavior.[11]
However, clinicians in the addictions field have postulated that opiates, such as methadone, have an immunosuppressant effect that
decreases host resistance to the AIDS virus,[12] particularly if there is
a pattern of multiple drug abuse involving alcohol, which is often
seen in drug maintenance programs. At high doses, alcohol is an
immunosuppressant.[13] If, in fact, the drug maintenance approach is
associated with a high incidence of multiple drug abuse patterns,
and if these drugs are themselves immunosuppressants, then it is
likely that a massive expansion of drug maintenance programs will
not reduce the morbidity and mortality associated with AIDS.[14] Furthermore, in association with the recent upswing in cocaine abuse,
there is an increase of IV cocaine injection in individuals on methadone maintenance, and inasmuch as cocaine is not blocked by
methadone, the drug maintenance approach may be less effective in
reducing needle use than has been assumed by drug maintenance
advocates.

Ryan and Pohl[15] suggested that infection with HIV alone is not
sufficient to cause an individual to develop AIDS. There may be
certain cofactors that team up to cause damage to the immune system along with HIV. Perhaps the most common suspected cofactors
are those that damage the immune system, such as alcohol and cer-

tain other drugs. It is known that alcohol, marijuana, cocaine and butyl nitrate, and perhaps opiates and PCP, damage the body's immune system. It is hypothesized that this combination of substance use and exposure to the AIDS virus may encourage the onset of AIDS in infected individuals.

There is also a growing body of clinical experience that suggests that elements of recovery-oriented chemical dependency treatment philosophy—which emphasizes a drug-free philosophy as well as good nutrition and exercise—may, in fact, enhance the immune system. The emphasis of the recovery-oriented treatment model of chemical dependency may, in the long term, better serve both the treatment of chemical dependency and AIDS issues that are associated with the chemical dependency treatment field. In addition, nonpsychoactive, nonimmunosuppressant chemical adjuncts to recovery would be more consistent with a sobriety-oriented philosophy as well as lessening immune system suppression. For example, naltrexone is a long-acting opiate receptor antagonist that does not have a psychoactive effect and it blocks opportunistic use of opiates. Its application as a treatment approach is compatible with recovery that is defined as learning to live a comfortable and responsible life without the use of psychoactive drugs.

There is some recent research that suggests that naltrexone may, in fact, enhance the endogenous endorphin system.[17] This enhancement of the endorphin system may interact with endorphin receptors on lymphocytes to enhance the immune system. Therefore, use of naltrexone as part of a recovery-oriented treatment philosophy would be more compatible with immune system enhancement than maintenance approaches employing psychoactive drugs, particularly when naltrexone is used as part of a full program of recovery.

However, concern has been expressed that addicts will not accept naltrexone in large enough numbers to have a major impact on slowing the spread of the AIDS epidemic. Clearly, the AIDS epidemic is having a major impact on chemical dependence treatment policy, but further study is needed to resolve the debate on drug maintenance versus drug-free treatment as well as the immunosuppressant effects of the maintenance drugs in combination with the various drugs of abuse that are involved in multiple drug abuse patterns.

DISTRIBUTION OF FREE NEEDLES
AND NEEDLE-SHARING RITUALS

Although it is believed that exposure to HIV and infection rates may be secondary to the virus itself in the shared needle and works, there is also a risk of immune damage that is secondary to the drugs themselves. However, needle sharing is an integral part of the drug-taking ritual, and sterile needles will not eliminate that ritual, but it will reduce some of the associated medical risk. Only those strategies that remove the individual from the IV drug abuse pool and provide for the elimination of immunosuppressant drugs will have the best chance of succeeding.

The assumption that distribution of free needles will change the ritual and that each individual will use their own outfit except when they are really high or junk sick is questionable at best, because addiction is characterized by a compulsion to use, the loss of control and continued use in spite of adverse consequences. In one study, 68% of the IV drug-using group engaged in needle sharing and said that they did so during 40% of their drug use episodes. Among the needle-sharing group, most (77%) said that they shared only with relatives and close friends, while others (32%) reported sharing needles with casual acquaintances and strangers. A variety of drugs was used by the study group and needle sharing was reported by over 59% of the men in each primary drug group.[18]

NEEDLE SHARING AND AIDS

A study in Dallas confirmed that needle sharing is common and frequently practiced among IVDUs. Other reports, however, have indicated that street addicts are aware of the risk of AIDS associated with needle sharing and show some interest in using new or clean needles.[19] The fact that the majority of needle-sharing individuals in the Dallas group did so only with others who were well known to them indicates that IVDUs are exercising some caution. This trend may indicate that IVDUs may be responsive to public information campaigns about the hazards of needle sharing.

The researchers in the Dallas study warned that "merely increasing the availability of needles without undertaking educational intervention might have a limited effect." They maintained that needle sharing is not only due to the shortage of needles but has also been found "to be associated with socialization, communal feeling, and protection in the drug culture."[20] (See Chart 1.)

Needle exchange programs in which the addict exchanges a "dirty outfit" for a "clean outfit" are being studied by both the California Medical Association and the California Society for the Treatment of Alcoholism and Other Drug Dependencies as a way of removing contaminated outfits from the drug culture, while at the same time increasing contact with the health care delivery system.

SEX EDUCATION AND CHEMICAL DEPENDENCY

Sex education, including safer-sex practices, should be instituted in chemical dependency treatment programs. It is important that programs include sex education and AIDS education for both staff and clients. Such educational approaches may conflict with homophobic, AIDS-phobic and sex-negative attitudes on the part of the treatment staff. In addition, religious-community and/or family

Chart I - Incidence of Needle Sharing

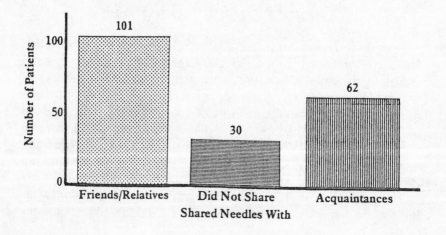

opposition to sex education and safer-sex guidelines — particularly in adolescent treatment programs — may impair the staff's ability to realistically deal with sexual issues in treatment programs.

In San Francisco, the highest risk adolescents are the homeless teenagers who live on the street. Many inject drugs, have multiple sexual partners, and prostitute themselves to gay and bisexual males. Young male prostitutes who are IVDUs are probably the highest risk group, but female prostitutes who are IVDUs also represent a bridge to the spread of AIDS into other populations.[21]

Expansion of AIDS education to adolescents is not only crucial in the schools and in chemical dependency treatment programs, but it also must extend to community outreach programs that deal with street adolescents. Denial of appropriate drug and sex education to young people for political, religious and ideological reasons only increases the possibility that AIDS will spread even more rapidly in this age group.

It may also be necessary to implement staff support groups in order to deal with AIDS phobia in high-risk programs as part of the ongoing staff training and client education program.[22] For example, the need has been clearly established for specific techniques in counseling gay male substances abusers who are at risk for AIDS.[23]

AIDS AND CHEMICAL DEPENDENCY: EDUCATION, COUNSELING, AND ANTIBODY TESTING

Since HIV is transmitted sexually as well as through the exchange of blood, it is extremely important to know the recommended guidelines for safer sex in order to reduce the risk of passing infection or becoming infected. Sexually active gay or bisexual men — especially those with a history of IV drug abuse — or heterosexual IV drug abusers who have shared needles or their sexual partners have a higher probability of already having been exposed to the AIDS virus. They may either carry the virus in their system or carry antibodies to the virus, which can be identified by the HIV antibody test. AIDS education is now identifying risk *behaviors* as opposed to risk groups, and it is these behaviors that place individuals at risk for infection with HIV. Thus, *any* sexually active individual with a number of *different* sexual partners must be informed of AIDS transmission and risk-reduction behaviors.

Ryan and Pohl have developed guidelines for education, counseling and antibody testing that can be used in chemical dependency treatment programs, stressing that it is critical for individuals with the disease of chemical dependency and alcoholism to be well informed about AIDS. It is particularly essential for gay men to realize that a gay male drug abuser, especially if he is an IVDU, is at the highest risk for having been exposed to the AIDS virus.

Testing Recommendations

The HIV antibody test was made available for general use in March 1985, and since that time a nationwide network "alternative" of test sites has developed to provide anonymous and confidential testing. Often, testing is available free of charge or for a small donation. No records are kept at these sites of an individual's test results. However, this is not the case for HIV antibody testing provided in hospitals, clinics, doctors' offices or other settings where test results are entered into patients' medical records.

Because of growing concern that positive test results might be used to discriminate against groups of individuals in regard to employment, housing, life and health insurance, child custody and other areas of civil rights, many health care providers are recommending that individuals at risk for AIDS who choose to get the antibody test do so at alternative test sites. In those states where mandatory reporting laws exist for anyone who tests positive, individuals obtaining the test have often done so under assumed names.[25] Pascal has reviewed the complicated legal issues surrounding the HIV antibody test and AIDS in drug treatment programs, and he emphasized that professionals need to be aware of the various confidentiality statutes.[26]

Procedure For Counseling the HIV Antibody Positive Individual

Counseling the seropositive individual and his/her significant other is an extremely important clinical issue. Ryan and Pohl[27] have developed the following guidelines (which have been synopsized for the purposes of this article) for counseling antibody positive individuals.

First, assess what the individual knows about the test, either as a result of prior information or pretest counseling, and then explain what the test means in simple, comprehensible language. Then assess the risk factors and inform the individual that this information is important to determine their health status and whether or not they are at risk for HIV infection. All information that is provided about the individual's risk status should be delivered in open, nonjudgmental terms.

If the individual is at low or no perceived risk for possible infection because there was no clear exposure to HIV, they should be informed of such. An important caveat, however, is that it is sometimes difficult to assess an individual's true risk status because they may not always have been completely truthful about, or actually remember (as in the case of drug blackouts), their behaviors. This is important to bear in mind because some people will have a positive test result even though they are not infected with the virus. The proportion of false positives is higher in low-risk groups who have no history of possible exposure to symptoms. Unfortunately, the test only reads antibodies to HIV and it cannot distinguish between which individuals are actually carrying the virus and which are not.

When informing an individual of a positive test result it is important to watch for shock or delayed emotional reactions to this information. At this time, the counselor must explain the range of responses that a positive test result can mean, such as a false positive; or that there has been exposure to the virus but the immune system has fought it off and only the antibodies remain; or that there has been exposure to the virus, antibodies have developed and the virus is present, but that they will probably not develop symptoms and may never become ill; or that they have developed antibodies, are carrying the virus and have a 10 to 40% chance of developing AIDS over the next 10 years. (It is very important to be careful with depressed or otherwise mentally or emotionally disordered patients.) Also, the individual should be given an opportunity to respond to the test results and ask questions, or if necessary the counselor should try to draw them out to determine their emotional response to the test results. Then the individual should be informed that they will be referred to a physician for additional medical follow-up; such a referral should be readily at hand as part of the primary aftercare treatment plan. Lastly, the individual needs to know that the majority of people exposed to the virus will not develop AIDS, but that they must be very careful to follow the recommended guidelines because they are a carrier of the virus.

Confidentiality

One last issue that counselors need to address is that of confidentiality. Confidentiality is the basis for treatment within chemical dependency treatment programs, and there is concern about the violation of confidentiality in treatment programs versus pressure to violate confidentiality because of the AIDS epidemic. AIDS treatment professionals and many chemical dependency treatment professionals support reporting AIDS as a communicable disease even if that violates confidentiality guidelines within chemical dependency treatment programs. Part of the controversy revolves around the concept of AIDS as a reportable illness in most jurisdictions. However, HIV positivity is another story.

Within 12-Step recovery-oriented community treatment programs there is a much greater emphasis on confidentiality and anonymity because of the tradition of anonymity in 12-Step programs and fellowships.[28] Major violations in confidentiality and anonymity will seriously impair the growing link between chemical dependency treatment and the 12-Step recovery network. Furthermore, there is great concern that mandatory reporting and violation of confidentiality in regard to AIDS will impair outreach approaches to chemically dependent clients, making them less likely to seek treatment early in their disease.

It should be noted that the American Medical Society on Alcoholism and Other Drug Dependencies — as well as other groups who are combatting drug dependency — has developed guidelines for facilities that treat chemically dependent patients who are at-risk for AIDS and who are infected by the HIV virus.[29]

CONCLUSION

Now and in the immediate future, substance abuse treatment professionals will play a major and expanding role in the AIDS epidemic, and it is crucial that chemical dependency treatment programs study both individual and organizational guidelines so that they can be implemented in a manner that is compatible with the treatment of addictive disease, which is itself a major public health problem. If treatment programs do not develop an AIDS and chemical dependency prevention and treatment component, chemical dependency treatment may be significantly altered in such a way as to make it less effective. AIDS and chemical dependency is a multidimensional problem that requires a multidisciplinary response. Treatment professionals need to regularly and continuously update themselves in the latest AIDS and chemical dependency guidelines in education, medical management and counseling. Also, chemical dependency treatment programs in both the public and private sectors, including adolescent treatment programs, need to be more active in implementing AIDS and chemical dependency guidelines in their programs.

Further research is needed to determine if the assumptions underlying AIDS and chemical dependency policy recommendations (e.g., needle sharing and drug maintenance) are in fact valid. Finally, chemical dependency treatment professionals need to play a greater role in determining public policy in regard to AIDS and chemical dependency treatment, as the AIDS epidemic will increasingly dictate public chemical dependency treatment policy.

NOTES

1. Centers for Disease Control. Heterosexual transmission of human t-lymphotrophic virus type III/lymphadenopathy-associated virus. Mortality and Morbidity Weekly Report. 1985; 34:245-248.

2. New York City Department of Health, Surveillance Office. AIDS surveillance update. 1985; Oct. 30.

3. Smith DE. et al. PCP and sexual dysfunction. In: Smith DE, Wesson DR, Buxton ME, Seymour RB, Ross S, Bishop MP & Zerkin EL. eds. PCP: Problems and Prevention. Dubuque, Iowa: Kendall/Hunt, 1982.

4. Abrams DI. The nature of AIDS. In: Acquired Immune Deficiency Syn-

drome and Chemical Dependency. Washington, D.C.: National Institute on Alcohol Abuse and Alcoholism, 1987.

5. Ostrow DC. Barriers to the recognition of links between drug and alcohol abuse and AIDS. In: Acquired Immune Deficiency Syndrome and Chemical Dependency. Washington, D.C.: National Institute on Alcohol Abuse and Alcoholism, 1987.

6. Selwyn PA. et al. Knowledge about AIDS and high-risk behavior among intravenous drug abusers in New York City. Paper presented at annual meeting of the American Public Health Association, Washington, D.C., 1985.

7. Smith DE. et al. op. cit.

8. Newmeyer, JA, Feldman HW, Biernacki P & Watters JK. Preventing AIDS contagion among intravenous drug abusers in New York City. Unpublished manuscript, 1986.

9. Griffin KM. Bleach: New weapon vs. AIDS. Am Med News. 1987; 1:21-22.

10. McConnell H. AIDS must supercede other drug issues. The Journal (of Addiction Research Foundation). 1986; 15(12): 1-2.

11. Selwyn PA. et al. op. cit.

12. Gil-Ad I, Bar-Yoseph J, Smadja Y, Zohar M & Laron Z. Effect of clonidine on plasma beta-endorphin, cortisol and growth hormone secretion in opiate-addicted subjects. Israel Journal of Medical Sciences. 1985; 21: 601-604.

13. Siegel L. AIDS: Relationship to alcohol and other drugs. J Substance Abuse Treatment. 1986; 3: 271-274.

14. MacGregor RR. Alcohol and the immune system. In: Acquired Immune Deficiency Syndrome and Chemical Dependency. Washington, D.C.: National Institute on Alcohol Abuse and Alcoholism, 1987.

15. Ryan C & Pohl M. Protocol for AIDS education and risk reduction counseling in chemical dependency treatment settings. Waltham, Massachusetts: ARC Research Foundation, 1986.

16. Siegel L. op. cit.

17. Wesson DR. Personal communication. Unpublished data. 1987.

18. Black JL, Dolan MD, DeFord HA, Rubenstein JA, Penk WE, Robinowitz R & Skinner BS. Sharing of needles among users of intravenous drugs. N Eng J Med. 1986; 314(7): 446-447.

19. Howard J & Borges P. Needle sharing in the Haight: Some social and psychological functions. J Health Soc Behav. 1970; 11(3): 220-230.

20. New study unveils sharing practices. AIDS and needle sharing. Street Pharmacologist. 1986; XI(4): 1.

21. Selwyn PA. AIDS. What is now known. New York: HP Publishing, 1986.

22. Smith DE. et al. op cit.

23. Smith TM. Counseling gay men about substance abuse and AIDS. In: Acquired Immune Deficiency Syndrome and Chemical Dependency. Washington, D.C.: National Institute on Alcohol Abuse and Alcoholism, 1987.

24. Ryan C & Pohl M. op. cit.

25. Ryan C & Pohl M. op. cit.

26. Pascal CB. Selected legal issues about AIDS for drug abuse treatment programs. J Psychoactive Drugs. 1987; 19(1): 1-12.

27. Ryan C & Pohl M. op. cit.

28. Buxton ME, Smith DE & Seymour RB. Spirituality and other points of resistance to the 12-Step recovery process. J Psychoactive Drugs. 1987; 19(3): 275-286.

29. American Medical Society on Alcoholism and Other Drug Dependencies. Guidelines for facilities treating chemically dependent patients at risk for AIDS or infected by HIV virus. New York: American Medical Society on Alcoholism and Other Drug Dependencies, 1987.

SELECTIVE GUIDE TO CURRENT REFERENCE SOURCES ON TOPICS DISCUSSED IN THIS ISSUE

Acquired Immune Deficiency Syndrome and Chemical Dependency

James E. Raper, Jr, MSLS
Lynn Kasner Morgan, MLS

Each issue of *Advances in Alcohol and Substance Abuse* features a section offering suggestions on where to look for further information on included topics. In this issue, our intent is to guide readers to selected sources of current information on acquired immune deficiency syndrome and chemical dependency.

Some reference sources utilize designated terminology (controlled vocabularies) which must be used to find material on topics of interest. For these a sample of available search terms has been indicated to assist the reader in accessing suitable sources for his/her purposes. Other reference tools use keywords or free-text terms (generally from the title of the document, the abstract, and the name

The authors are affiliated with the Gustave L. and Janet W. Levy Library, The Mount Sinai Medical Center, Inc., One Gustave L. Levy Place, New York, NY 10029-6574.

of the responsible agency or conference). In searching the latter, the user should also look under synonyms for the concept in question.

An asterisk (*) appearing before a published source indicates that all or part of that source is in machine-readable form and can be accessed through an online database search. Database searching is recommended for retrieving sources of information that coordinate multiple concepts or subject areas. Most health sciences libraries offer database services, and many databases are now available for searching in one's office or home via subscriptions with database vendors and a microcomputer equipped with a modem.

Of particular relevance to this issue, the National Library of Medicine publishes recurring bibliographies on acquired immuno-deficiency syndrome (AIDS), which contain references to articles from some 3,000 journals and a limited number of books, at present covering January 1980 through June 1987. Citations to all preclinical, clinical, epidemiologic, diagnostic, and preventive areas are included. These bibliographies are available free of charge from the Literature Search Program, Reference Section, National Library of Medicine, 8600 Rockville Pike, Bethesda, Md. 20894.

The following government agencies provide AIDS hotline informational and referral services for patients and physicians:

Centers for Disease Control
1600 Clifton Road, NE
Atlanta, Georgia 30333
1-800-342-7514

National Cancer Institute
Office of Cancer Communications
9000 Rockville Pike
Bethesda, Md. 20892
1-800-422-6237

In addition, there are many other national, state, and local organizations that provide AIDS and substance dependence information and services.

Readers are encouraged to consult their librarians for further assistance before undertaking research on a topic.

Suggestions regarding the content and organization of this section are welcome.

1. INDEXING AND ABSTRACTING SOURCES

Place of publication, publisher, start date, and frequency of publication are noted.

AIDS (Acquired Immune Deficiency Syndrome). Phoenix, Ariz., Oryx Press, v.1, 1985- , annual.
    See: Table of Contents.
*Biological Abstracts (1926-   ) and *Biological Abstracts/RRM* (v.18, 1980-   ). Philadelphia, BioSciences Information Service, semimonthly.
    See: Biosystemic index.
    See: Generic index.
    See: Keyword-in-context subject index.
*Chemical Abstracts. Columbus, Ohio, American Chemical Society, 1907- , weekly.
    See: *Index Guide* for cross-referencing and indexing policies.
    See: *General Subject Index* terms, such as alcoholic beverages, drug dependence, drug-drug interactions, drug tolerance, immunosuppression.
    See: Keyword subject indexes.
*Dissertation Abstracts International. Section B. The Sciences and Engineering. Ann Arbor, Mich., University Microfilms, v.30, 1969/70- , monthly.
    See: Keyword-in-context subject index.
*Excerpta Medica: Adverse Reactions Titles. Section 38. Amsterdam, The Netherlands, Excerpta Medica, v.15, 1980- , monthly.
    See: Subject index.
*Excerpta Medica: Clinical Biochemistry. Section 29. Amsterdam, The Netherlands, Excerpta Medica, v.27, 1973- , 32 issues per year.
    See: Subject index.
*Excerpta Medica: Drug Dependence. Section 40. Amsterdam, The Netherlands, Excerpta Medica, v.8, 1980- , 6 issues per year.

See: Subject index.

Excerpta Medica: Drug Literature Index. Section 37. Amsterdam, The Netherlands, Excerpta Medica, v.7, 1975- , 24 issues per year.

See: Subject index.

Excerpta Medica: Immunology, Serology and Transplantation. Section 26. Amsterdam, The Netherlands, Excerpta Medica, 1967- , 30 issues per year.

See: Subject index.

Excerpta Medica: Internal Medicine. Section 6. Amsterdam, The Netherlands, Excerpta Medica, 1947- , 30 issues per year.

See: Subject index.

Excerpta Medica: Pharmacology. Section 30. Amsterdam, The Netherlands, Excerpta Medica, v.57, 1983- , 20 issues per year.

See: Alcoholism, drug addiction sections.

See: Subject index.

Excerpta Medica: Psychiatry. Section 32. Amsterdam, The Netherlands, Excerpta Medica, v.22, 1969- , 20 issues per year.

See: Addiction, alcoholism sections.

See: Subject index.

Excerpta Medica: Public Health, Social Medicine and Hygiene. Section 17. Amsterdam, The Netherlands, Excerpta Medica, 1955- , 20 issues per year.

See: Addiction, drug control sections.

See: Subject index.

Excerpta Medica: Toxicology. Section 52. Amsterdam, The Netherlands, Excerpta Medica, 1983- , 20 issues per year.

See: Subject index.

Index Medicus. (Including *Bibliography of Medical Reviews*). Bethesda, Md., National Library of Medicine, 1960-, monthly.

See: *MeSH* terms, such as acquired immunodeficiency syndrome, alcoholism, drug interactions, HIV, human T-Cell leukemia virus, narcotics, substance dependence, substance use disorders, pharmacology, toxicology.

Index to Scientific Reviews. Philadelphia, Institute for Scientific Information, 1974- , semiannual.

See: Permuterm keyword subject index.

See: Citation index.

International Pharmaceutical Abstracts. Washington, D.C., American Society of Hospital Pharmacists, 1964- , semimonthly.

> See: IPA subject terms, such as acquired immunodeficiency syndrome, alcoholism, dependence, drug abuse.

Psychological Abstracts. Washington, D.C., American Psychological Association, 1927- , monthly.

> See: Index terms, such as addiction, drug abuse, drug addiction, drug dependency, drug usage, immunologic disorders.

Public Affairs Information Service Bulletin. New York, Public Affairs Information Service, v.55, 1969- , semimonthly.

> See: PAIS subject headings, such as acquired immunodeficiency syndrome, alcoholism, drug abuse, drug addicts.

Science Citation Index. Philadelphia, Institute for Scientific Information, 1961-, bimonthly.

> See: Permuterm keyword subject index.
> See: Citation index.

Social Work Research and Abstracts. New York, National Association of Social Workers, v.13, 1977- , quarterly.

> See: Fields of service sections, such as alcoholism and drug addiction.
> See: Subject index.

2. CURRENT AWARENESS PUBLICATIONS

Current Contents: Clinical Practice. Philadelphia, Institute for Scientific Information, 1973- , weekly.

> See: Keyword index.

Current Contents: Life Sciences. Philadelphia, Institute for Scientific Information, v.10, 1967- , weekly.

> See: Keyword index.

Current Contents: Social and Behavioral Sciences. Philadelphia, Institute for Scientific Information, v.6, 1974- , weekly.

> See: Keyword index.

3. BOOKS

Andrews, Theodora. *A Bibliography of Drug Abuse, Including Alcohol and Tobacco*. Littleton, Colo., Libraries Unlimited, 1977- .

Andrews, Theodora. *Guide to the Literature of Pharmacy and the Pharmaceutical Sciences*. Littleton, Colo., Libraries Unlimited, 1986.

Medical and Health Care Books and Serials in Print: An Index to Literature in the Health Sciences. New York, R. R. Bowker Co., annual.

    See: Library of Congress subject headings, such as acquired immunodeficiency syndrome, alcoholics, alcoholism, drug abuse, narcotic habit, pharmacology, toxicology.

National Library of Medicine Current Catalog. Bethesda, Md., National Library of Medicine, 1966- , quarterly.

    See: *MeSH* terms as noted in Section 1 under *Index Medicus*.

4. U.S. GOVERNMENT PUBLICATIONS

Monthly Catalog of United States Government Publications. Washington, D.C., U.S. Government Printing Office, 1895- , monthly.

    See: Following agencies: Alcohol, Drug Abuse and Mental Health Administration, Centers for Disease Control, National Cancer Institute, National Institute of Mental Health, National Institute on Drug Abuse.

    See: Subject headings, derived chiefly from the Library of Congress, such as acquired immunodeficiency syndrome, alcoholism, drug abuse, drug interactions, narcotics, pharmacology, toxicology.

    See: Keyword title index.

5. ONLINE BIBLIOGRAPHIC DATABASES

Only those databases which have no single print equivalents are included in this section. Print sources which have online database equivalents are noted throughout this guide by the asterisk (*)

which appears before the title. If you do not have direct access to these databases, consult your librarian for assistance.

AIDS ABSTRACTS (Bureau of Hygiene and Tropical Diseases, London, England).
   Use: Keywords.
ASI: AMERICAN STATISTICS INDEX (Congressional Information Services, Inc., Washington, D.C.)
   Use: Keywords.
DRUG INFO/ALCOHOL USE/ABUSE (Hazelden Foundation, Center City, Minn., and Drug Information Service Center, College of Pharmacy, University of Minnesota, Minneapolis, Minn.).
   Use: Keywords.
MAGAZINE INDEX (Information Access Co., Belmont, Calif.).
   Use: Keywords.
MEDICAL AND PSYCHOLOGICAL PREVIEWS: MPPS (BRS Bibliographic Retrieval Services, Inc., Latham, N.Y.; formerly *PRE-MED* and *PRE-PSYCH*).
   Use: Keywords.
MENTAL HEALTH ABSTRACTS (IFI/Plenum Data Co., Alexandria, Va.).
   Use: Keywords.
NATIONAL NEWSPAPER INDEX (Information Access Co., Belmont, Calif.).
   Use: Keywords.
NTIS (National Technical Information Service, U.S. Dept. of Commerce, Springfield, Va.).
   Use: Keywords.
PDQ (National Library of Medicine and National Cancer Institute, Bethesda, Md.)
   Use: Cancer information, which includes acquired immune
        deficiency syndrome (AIDS), and other menu options.
PSYCALERT (American Psychological Association, Washington, D.C.)
   Use: Keywords.

6. HANDBOOKS, DIRECTORIES, GRANT SOURCES, ETC.

Annual Register of Grant Support. Wilmette, Ill, National Register Pub. Co., annual.
> See: Medicine; pharmacology; psychiatry, psychology, mental health sections.
> See: Subject index.

*Encyclopedia of Associations. Detroit, Gale Research Co., annual (occasional supplements between editions).
> See: Subject index.

*Foundation Directory. New York, The Foundation Center, biennial (updated between editions by *Foundation Directory Supplement*).
> See: Index of foundations.
> See: Index of foundations by state and city.
> See: Index of donors, trustees, and administrators.
> See: Index of fields of interest.

Research Awards Index. Bethesda, Md., National Institutes of Health, Division of Research Grants, annual.
> See: Subject index.

O'Brien, Robert and Sidney Cohen. *The Encyclopedia of Drug Abuse*. New York, Facts on File Pub., 1984.

7. JOURNAL LISTINGS

*Ulrich's International Periodicals Directory. New York, R. R. Bowker Co., annual (updated between editions by *Ulrich's Quarterly*).
> See: Subject categories, such as drug abuse and alcoholism, medical sciences, pharmacy and pharmacology, psychology.

8. AUDIOVISUAL PROGRAMS

*National Library of Medicine Audiovisuals Catalog. Bethesda, Md., National Library of Medicine, 1977- , quarterly.
> See: *MeSH* terms as noted in Section 1 under *Index Medicus*.

Patient Education Sourcebook. [Saint Louis, Mo.], Health Sciences Communications Association, c1985.
    See: *MeSH* terms as noted in Section 1 under *Index Medicus*.

9. GUIDES TO UPCOMING MEETINGS

Scientific Meetings. San Diego, Calif., Scientific Meetings Publications, quarterly.
    See: Subject indexes.
    See: Association listing.
World Meetings: Medicine. New York, Macmillan Pub. Co., quarterly.
    See: Keyword index.
    See: Sponsor directory and index.
World Meetings: Outside United States and Canada. New York, Macmillan Pub. Co., quarterly.
    See: Keyword index.
    See: Sponsor directory and index.
World Meetings: United States and Canada. New York, Macmillan Pub. Co., quarterly.
    See: Keyword index.
    See: Sponsor directory and index.

10. PROCEEDINGS OF MEETINGS

Conference Papers Index. Louisville, Ky., Data Courier, v.6, 1978- , monthly.
Directory of Published Proceedings. Series SEMT. Science/Engineering/Medicine/Technology. White Plains, N.Y., InterDok Corp., v.3, 1967- , monthly, except July-August, with annual cumulations.
Index to Scientific and Technical Proceedings. Philadelphia, Institute for Scientific Information, 1978- , monthly with semiannual cumulations.

11. SPECIALIZED RESEARCH CENTERS

International Research Centers Directory. 4th ed. Detroit, Gale Research Co., 1988-89, c1988.

Research Centers Directory. 11th ed. Detroit, Gale Research Co., 1987 (updated by *New Research Centers*).

12. SPECIAL LIBRARY COLLECTIONS

Ash, L., comp. *Subject Collections*. 6th ed. New York, R. R. Bowker Co., 1985.

Directory of Special Libraries and Information Centers. 10th ed. Detroit, Gale Research Co., 1987 (updated by *New Special Libraries*).